W0106984

SPRINGER
LABORATORY

A. Askar   H. Treptow

Quality Assurance in Tropical Fruit Processing

With 23 Figures and 36 Tables

Springer-Verlag
Berlin Heidelberg New York London Paris
Tokyo Hong Kong Barcelona Budapest

Professor Dr.-Ing. AHMED ASKAR

Department of Food Sciences
Faculty of Agriculture
Suez Canal University
Ismailia, Egypt

Dr.-Ing. HANS TREPTOW

TU Berlin
Institut für Lebensmitteltechnologie
Königin-Luise-Straße 22
1000 Berlin 33, Germany

ISBN-13:978-3-642-77689-2   e-ISBN-13:978-3-642-77687-8
DOI: 10.1007/978-3-642-77687-8

Library of Congress Cataloging-in-Publication Data. Askar, Ahmed. Quality assurance in tropical fruit processing / A. Askar, H. Treptow. p. cm. – (Springer laboratory) Includes index. ISBN-13:978-3-642-77689-2 1. Tropical fruit. 2. Tropical fruit industry–Quality control. I. Treptow, Hans. II. Title. III. Series. TP440.A85 1993 664'. 8046–dc20   93-9518

© Springer-Verlag Berlin Heidelberg 1993
Softcover reprint of the hardcover 1st edition 1993

Data conversion: Renner Typografie & Satz, 6901 Wiesenbach
52/3145- 5 4 3 2 1 0 – Printed on acid-free paper

*This book is dedicated
to our families*

Preface

Tropical and subtropical countries have become well aware of the fact, that they must make better use of their fruits. In spite of the favourable climatic conditions for the production of varieties of delicious fruits in such countries, continuously high temperatures shorten the shelf-life of most fruits and fruit products.

A tropical climate provides ideal conditions for rapid growth of spoilage microorganisms and for chemical reactions. Most of such reactions in fruits and fruit products are deteriorative in nature causing high respiration rates, texture softening and spoilage of fruit. This causes loss of colour, flavour and vitamins, and browning of fruit products. Even though a fruit product has been rendered microbiologically stable, these chemical reactions continue to occur in storage, and they occur much more rapidly in a tropical climate.

The processing of fruits and soft drinks is a predominate food industry in tropical and subtropical countries. Some of the large companies in such industries are partly foreign owned. They seem to be efficiently operated with adequate capital, good management, and technological competence, all of which are usually imported from the parent company. However, most of small and medium companies are locally owned, and are deficient in technology and management ability. The products are generally fair. It is rare to find a trained quality assurance manager in these companies. Processing of good fruit products, especially for export, requires sound fruit processing lines as well as good management that achieves internationally accepted standards of quality.

Departments of Food Technology should maintain a close relationship with the food industry and study both undergraduate and graduate student research on some of the formidable problems that the fruit processing industry faces in tropical and subtropical countries. It is hoped that this MANUAL will be useful in providing information for the growing tropical fruit industry in such regions.

The expected users of the MANUAL will include quality assurance managers and coworkers, fruit plant managers, product research development specialists, fruit product brokers, technical salesmen, food equipment manufacturers and fruit industry suppliers. The MANUAL is also designed to be used by undergraduate and graduate students and practising scientists in universities, research institutes and governmental control autorities.

In the MANUAL both the classical laborious methods of chemical, physical, microbiological and organoleptical method of analysis and the modern, sophisticated and rapid procedures of testing and analyzing are presented. Special attention is given to methods which are economical, give accurate results, and are rapid and easy to use. In this MANUAL a reader will find all the information on a particular test presented together: the principle, the purpose, chemicals and instruments involved and employed, preparation, and interpretation of results.

After an introductory chapter on „Quality Assurance Management", chapter 2 deals with the classical and modern analytical methods for testing fruit quality. Chapter 3 describes the physical measurements of colour, viscosity, texture and water activity. Information about sensory analysis, including statistical test designs and tables, is provided in chapter 4. In chapter 5 the major microbiological examinations of fruit products are given, followed by chemical tests used today for the detection of microbial contamination and spoilage. Chapter 6 summarizes the importance and standards of water used in this industry and describes the physical, chemical and microbiological tests used for quality control of water. A glossary of terms related to sanitation, factors affecting cleaning efficiency and costs and evaluation of cleanliness and sanitation in fruit processing plants is given in chapter 7. In addition, information about control of employee hygiene practices and of pest control are reviewed. The last chapter deals with terminology used in waste disposal control and summarizes factors to be considered in waste disposal. This final chapter also describes the method of analysis used in the evaluation of waste water.

We are especially indebted to all the researchers and authors who have worked in and written about quality assurance of food products and on problems of the food industry in tropical countries (see recommended literature in the individual chapters and in the appendix). Acknowledgement is also due to many in the fruit industry, including supply firms and companies producing test materials and equipment, who have willingly supplied literature, technical information and illustrative material used in this MANUAL.

Our families have provided us with encouragement and warm support throughout this project. We also thank Mr. Wolfgang Stellmach for his help in the computer work of the manuscript.

And finally, the encouragement and assistance of Mr. Peter Enders and our publisher, Springer-Verlag, who believed in our project, are gratefully acknowledged.

                                                                    AHMED ASKAR
Ismailia/Berlin, February 1993   HANS TREPTOW

Contents

1 Quality Assurance Management

Quality assurance plays an important role in the maintenance of quality of processed fruit products at levels and tolerances acceptable to the consumer. It also assures compliance with government regulations, reducing the probability of spoilage, minimizing the cost of production and increasing the product value or saleability.

In a small plant, the quality assurance group may consist of a single individual. In a larger organization a special team of quality assurance personnel and a suitable laboratory are required. A quality assurance program can be started with a minimum of expenditure and expanded as the need arises.

Too often thoughts of quality assurance in a fruit processing plant bring visions of laboratories full of complicated expensive equipment. Whereas, in reality, the laboratory is only one facet of quality assurance, being in essence a place where the results of quality assurance in the plant are checked. The quality of the control supervisor (or the quality assurance manager) and his staff are the most important factor in the quality assurance program, not the laboratory.

Quality assurance management in a fruit processing plant may be discussed under three headings:

– Personnel requirements,
– Organization and function, and
– Layout and laboratory requirements.

Personnel Requirements

The first task is the designation of a quality assurance manager. The people who come into question must have a university education, preferably in food science and technology. They must have experience in fruit processing and in quality control with a comprehensive knowledge of the literature in the field of food analysis and food microbiology. They should be able to plan their work and to experiment on their own.

Quality assurance managers should not only be able to work systematically, but should also have the talent of making use of all knowledge at hand.

They must be open to ideas from all fields of food science. They should have the spirit of a pioneer, never willing to give up.

They must have sales ability, have the ability to speak the language of the industry, be able to cooperate with the production manager and other managers, be always alert, responsive to necessary changes and honest in reports, decisions and, above all, in analysis.

It is of advantage to know several languages. English is the most widely used language, knowledge of others, however would help for better understanding.

The degree of technical training required for the quality assurance staff is largely dependent on the size of the operation and on the routine laboratory work. Some of them must have a university degree in food analysis and food microbiology. For routine analysis the technical-school leaver may be the most practical type of person.

One of the duties of quality assurance managers is the training of their own personnel. A person with real interest in this type of work is essential and, of course, some technical background is highly desirable.

Organization and Function of the Quality Assurance Department

The relationships between the quality assurance department and other departments within the fruit processing plant is very important (Fig. 1). The quality assurance manager should provide a channel of communication with other departments, including purchasing and marketing, warehousing, maintenance and the production department in particular. The quality reports must be distributed among other departments.

Organization Principles

The quality assurance manager and department must have a direct line to the top management (Fig. 1) and must have their complete support. He/She must have

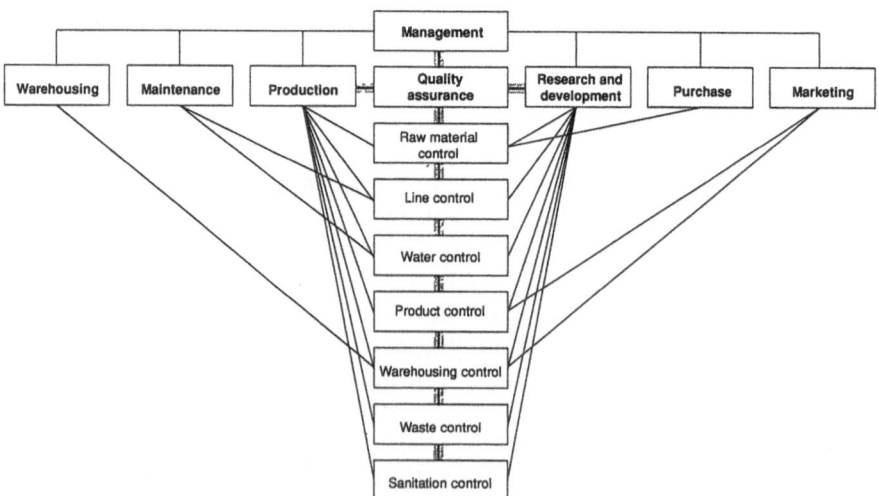

Fig. 1. Organization and function of a quality assurance department

the same status and authority as the production manager, and should be directly responsible to top management.

The quality assurance department should not be subordinated to the production department. *This is one law in the organization scheme that cannot be compromised.*

The responsibility of the production department is to achieve maximum quantities in terms of production yield and plant capacity. Whereas, the responsibility of the quality assurance department is to insure that this objective is achieved without impairment to the quality level of the plant and with the maximum profit.

Only the top management can make the decision between quality and quantity, not one of the several departments of the company.

The top manager must have the opinion of the quality assurance manager on all basic decisions, such as budget policies, sales policies, pricing, inventory, ... etc. In order for quality assurance to be more effective, a member of the quality assurance group (a quality assurance inspector), should be able to report directly to the production manager, who can take remedial steps immediately. The quality assurance inspector should note where and when remedial action is to be taken. If no action is taken by the production department, the quality assurance supervisor (or manager) should have the authority to discontinue the operation, or at least, to alert the top management of the failure to take action.

In small factories, the quality assurance department can be also responsible for research and development. Research and development, which must be done in every fruit processing plant, must have a close relationship with the quality assurance department. Research must be seen as an investment in the future.

Quality assurance must be considered as a science and not an art. It is not based on individual personal judgement, but on chemical, physical, microbiological and sensory tests treated statistically.

Function

Quality assurance should operate routinely in a continuous and close-co-operation with all of manufacturing process, starting with the buyer's specification to the sale of the finished product, without the need for detailed approval from the top management. However, it should never become routine work without improvement. The methodology has to be continuously developed.

The functions of the quality assurance department may be divided as follows (Fig. 1):

1. Inspection and grading of fruits as raw material (purchasing specifications), quality assurance of all food additives and other ingredients, and of packaging materials, through setting of specification and developing test procedures.

2. Quality assurance of processing operations (Line control), production efficiency, defining the critical points in the process.
3. Control of water, including process, boiler and cooling waters.
4. Quality assurance of the finished product, establishment of his own specifications, standards, shelf-life, and improvement of product quality.
5. Waste disposal control.
6. Warehousing control.
7. Sanitation control. The quality assurance manager should prepare a manual and every detail must be set down in writing. By the use of the manual as a guide, the function of quality assurance resolves itself into the periodic inspection and quality control.

Layout and laboratory requirements

It is essential that all fruit processing plants should have access to a laboratory equipped with the minimum instruments, apparatus and other allied facilities necessary to check the quality of raw material, food additives, water, finished products and sanitation.

The laboratory can be started with a minimum of expenditure and expanded as the need arises. It must be close to, but apart from the processing lines. For a medium size fruit processing plant a floor space of approximately 10 x 16 m is recommended (Fig. 2). It provides ample managerial office facilities (with 3 desks, filing cabinets, library), working space for chemical analysis, and sensory evaluation, and storage areas for retaining samples and chemicals.

Many other innovations and arrangements are possible, but the layout in Fig. 2, has been found to be quite satisfactory.

Benches should be equipped with electric power, water, gas, vacuum, and air pressure outlets, must be acid- and alkali-resistant, and must have a sink capable of heavy duty waste disposal. On the wall above the benches, shelves should be constructed for the storage of chemicals, samples and reagents used daily.

The room for sensory analysis should be constructed in such a way that the panel members can work individually when evaluating the samples (Chapter 4), provided with overhead lighting with various coloured lights. The walls should be pale green.

Equipment requirements will vary considerably depending on the product packed, size of the plant and quality of the product. The basic items for a medium size fruit processing plant include:

– Hand refractometers, Abbe refractrometer,
– Analytical balances, top loading balance 1 kg, 5 kg and 10 kg,
– Refrigerator, freezer, distilled water still,
– Digital single beam sprectrophotometer, pH-meter, thermometers,
– Drying oven to 200 C, vacuum oven 0.1 mm Hg),
– Muffle oven to 600 C or 1000 C,

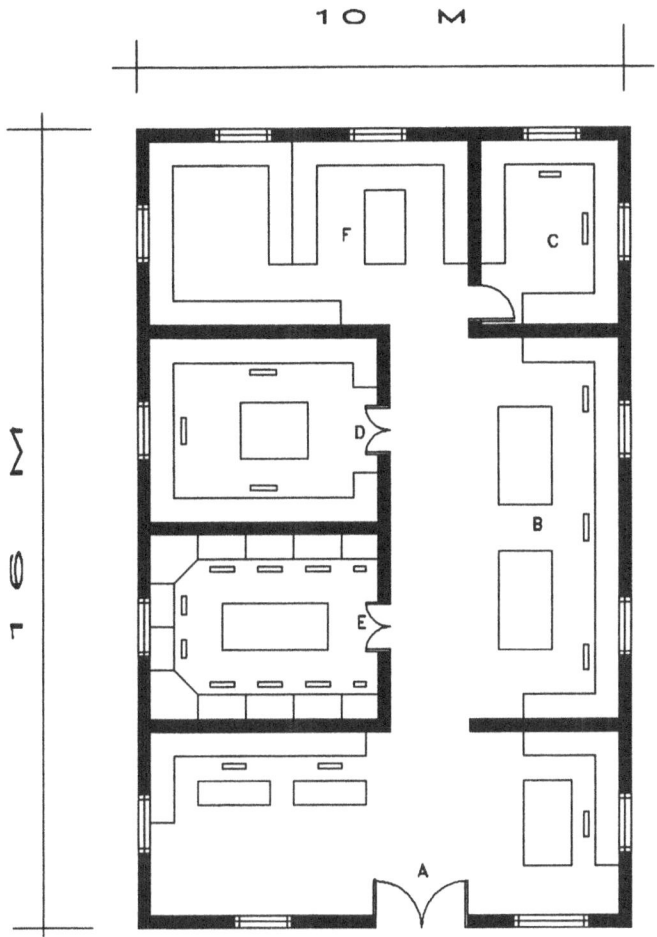

Fig. 2. Layout of quality assurance department A = Manager office, reception, desks, library, files, secretary; B = Chemical analysis; C = Physical analysis; D = Microbiological analysis; E = Sensory analysis and F = Storage for samples and chemicals

- Warring blender, automatic titrometer, hot plates, water baths,
- Rotary evaporator, high speed centrifuge, viscometer,
- Autoclave, incubators (37 and 55 C), laminar flow, colony counter.

2 Analytical Methods

Water in fruits and fruit products occurs in three different forms: "free water" as a solvent or dispersing medium, "absorbed water" on the surface of colloidal particles (proteins, polysaccharides ...) or as capillary water, and "bound water" of hydration in chemical combination and hydrates of various salts. The moisture content can be determined by different methods:

- Weight lost due to drying, the usual method.
- Distillation with a solvent having a higher boiling point and lower specific gravity than water, e.g. toluene (114 C, and 0.8669). This method is of special importance for fruit products containing volatile substances. The refluxed water settles as the solvent floats in a graduated tube, in which it can be measured by volume.
- The Karl Fischer method, which is based on the nonstoichiometric reaction of water with iodine and sulphur dioxide in pryridine solution.
- Physical methods using an Electrical Moisture Meter or based on the measurement of nuclear magnetic resonance absorption of hydrogen nuclei in water.

Determination of Moisture by Drying

- Vacuum oven at 70 ± 1 C and 100 mm Hg (not to exceed 450 mm Hg), Forced draft air oven 100–105 C, and /or top-loading balance with infrared heating element.
- Analytical balance
- Aluminium weighing dishes
- Desiccator

Spread a 5–10 g ground sample over the bottom of a previously dried and tared metal dish. After drying for 6 hours remove to a desiccator, cool and weigh. The drying time should be sufficient to give a constant mass. For most fruit products 6–10 hours is adequate.

The moisture content is usually expressed as percentage mass of water: original mass, but with dried fruits it is sometimes expressed as percentage mass of water: dry mass.

The drying in a vacuum oven is recommended, because of the decomposition of sugar and the casehardening in fruits and fruit products. During drying, a slow current of air admitted to the oven (ca. 2 bubbles/s) dried by passing through H_2SO_4. A subsequent determination should agree within 0.2%. Drying for 6 h is generally sufficient.

For the determination of moisture content in sugar, citric acid and other food additives a normal oven at 100–105 C for 2–4 hr can be used.

The proportion of free water lost increases as the temperature is raised, therefore it is especially important to compare only the results obtained using the same drying conditions. Particle size, weight of sample, type of dish and oven type may affect the results.

The moisture content can be also measured using an "Infrared Moisture Balance". This provides continual indication of weight decrease or moisture loss. The instrument is especially useful since weighing and drying are simultaneous.

Determination of Moisture by the Karl Fischer Method

Reagents:

a) Karl Fischer Reagent: Dissolve 133 g Iodine in 425 ml dry pyridine. Add 425 ml dry ethylene glycol monoethyl ether. Cool to 4 C in ice bath and bubble in 102–105 g SO_2. Mix well and allow to stand for 12 hours.

Place 50 ml formamide into a Berzelius beaker containing a magnetic stirrer. Place in titrimeter and titrate with the reagent to end point. Quickly add an accurately weighed amount of sodium tartrate dehydrate (0.25–0.35 g). Titrate immediately to same end point. Calculate the mg H_2O/ml reagent = [mg Sodium tartrate dehydrate x 0.1566] /ml reagent.

b) Sodium tartrate dehydrate 60 mesh.

c) N,N-dimethylformamide (DMF).

d) Formamide.

e) Pyridine.

f) Ethylene glycol monoethyl ether.

Determination: Accurately weigh 2 g ground sample into a 50 ml Erlenmeyer flask. Pipette in 20 ml DMF. Tape stopper securely to flask and heat 60 min at 35 C in oven. Shake flask 10 min and cool to room temperature. Decant the solution into a centrifuge tube and centrifuge at 1500 rpm for 5 min. Place 50 ml formamide into a Berzelius beaker and titrate to end point with Karl Fischer reagent. Quickly pipette 10 ml sample solution into beaker and titrate to the same end point. Carry out a titration of 10 ml DMF in the same manner as the sample.
Calculation:

$$a = \frac{[200\,(b-c)\,d]}{e}$$

Where:

a = % H_2O in sample
b = ml reagent for sample
c = blank titration
d = mg H_2O/ml reagent
e = mg sample

The Karl Fischer method is employed mainly for the determination in materials fairly low moisture content, e.g. dried fruits, sugar confectionery and instant powders. More information on this method is given by Zurcher and Hadorn [2].

2.2 Total Soluble Solids (Brix)

The concentration of soluble solids in a large volume of fruit juice may be determined by hydrometry or refractometry. Refractometry is an appropriate method if only small samples are available. Hydrometers are hollow glass "spindels" terminating at the lower end in a weighted bulb and having the upper end in the form of a slender stem within which a graduated scale is sealed. When floated in a juice, a hydrometer sinks to a depth determined by the specific gravity of the juice, which is related to the concentration of sugar and other soluble solids [3]. Brix (or Balling) hydrometers are calibrated to read directly the percentage by weight of sucrose in pure solutions of sucrose in water.

The measurement of soluble solids using a refractometer is the most usual test for routine control purposes. The instrument generally used for determining of total soluble solids (TSS) in fruit processing plants is either an Abbe or hand refractometer [4]. The refractometric method is also applicable to less fluid fruit products such as jams, and pulps that cannot be tested by hydrometry. Although refractometers differ in design, all use the critical angle of total reflection to measure refractive index. The observer sees an optical field partly obscured by a shadow with a sharp boundary, the position of which is determined by the refractive index of the sample.

If preliminary pulping of the sample is necessary, filter a small portion of the pulped sample through a pledget of absorbent cotton in a small funnel. Discard the first few drops, then place 2 or 3 drops on the refractometer prism. Determine the direct refractometer reading at 20 C.

It gives the refractive index as well as TSS (calculated as sucrose) at 20 C, since the refractometers are usually calibrated at 20 C (68 F). Temperature corrections of refractometers are given in Table 1.

The *Brix* value is at present the usual expression for TSS. Brix is defined as percent of sucrose measured by a Brix hydrometer, Brix and *Balling* are the same. Balling gives also the percent of sucrose by weight at the temperature indicated on the instrument.

The relation between percent of sucrose (Brix or Balling), specific gravity and refractive index are given in Table 2.

In the *Baumé* hydrometer the divisions range from 0 to 70. The original Baumé hydrometer scale is graduated so that for liquids heavier than water. 0 is the point to which the hydrometer sinks in water and 10 the point to which it sinks in a 10% solution of NaCl.

The relation between Brix-degree (or Balling-degree) and Baumé-degree of sucrose solutions are given in Table 3.

For samples containing invert sugar, such as fruits and its products, correct the percent solids by adding 0.022 for each per cent invert sugar in the sample.

Since fruit acids are the main non-sugar soluble solids in the fruit products, corrections to be applied to refractometer reading to get true brix-degree, Table 4.

The corrections were developed mainly for citrus products, which contain citric acid, but are also applicable to other products containing known amounts of citric acid.

2.3 Determination of Sugars

2.3.1 Determination of Reducing Sugars

Several methods for the determination of reducing sugars in fruits and fruit products have been developed and recommended. According to our experience, the Lane and Eynon method is simple, rapid and more suitable for routine analysis.

Invert sugar reduces the copper in Fehling's solution to insoluble cuprous oxide (red). The sugar content in the sample is estimated by determining the volume of the unknown sugar solution required to completely reduce a measured volume of Fehling's solution. The Lane and Eynon-method is one of the methods recommended by the AOAC [8], for the determination of reducing sugars in food.

Reagents:

- *Soxhlet modification of Fehling's solution:* prepared by mixing equal volumes of Fehling's solutions (a) and (b) immediately before use.
- *Fehling's solution (a):* dissolve 34.639 g of copper sulphate ($CuSO_4 . 5 H_2O$) in water, up to 500 ml, and filter through No. 4 Whatman paper.
- *Fehling's solution (b):* dissolve 173 g of Rochelle salt (potassium sodium tartrate, $KNa C_4H_4O_6 . 4 H_2O$) and 50 g NaOH in water up to 500 ml, let to stand 2 days and filter through asbestos.

- *Invert sugar standard solution (1%):* weigh 9.5 g pure sucrose into a 1-liter volumetric flask. Add 100 ml water and 5 ml HCl. Store 7 days at 15 C (or 3 days at 20–25 C) for inversion to take place. Dilute up to mark with water. This solution (acidified 1% invert sugar) is stable for several month.
- *Before use,* pipette 25 ml of the standard invert sugar solution into a 100 ml-volumetric flask and neutralize with ca. 1 N NaOH using phenolphthalein as indicator. Make up to mark with water (1 ml = 2.5 mg of invert sugar).
- *Methylene blue indicator:* 1 g in 100 ml water.
- *Neutral lead acetate solution (45%):* dissolve 225 g in water up to 500 ml.
- *Potassium oxalate solution (22%):* dissolve 110 g in water up to 500 ml.

Determination of the Exact Amount of Potassium Oxalate Necessary to Precipitate the Lead

An excess of lead acetate in the sugar solution will result in an error in the titration. To obtain this value, pipette 2 ml aliquote of the lead acetate solution into each of six 50 ml beakers containing 25 ml water. To the beakers, add 1.6, 1.7, 1.8, 1.9, 2.0 and 2.1 ml potassium oxalate solution, respectively. Filter each through a 41H Whatman paper and collect the filtrates in 50 ml conical flasks. To each of the filtrates, add a few drops of potassium oxalate solution. The correct amount of potassium oxalate required is the smallest amount which, when added to 2 ml of lead acetate solution, gives a negative test for lead in the filtrate. In the presence of lead, the filtrate gives a white precipitate with HCl. The equivalent volume should be marked on the bottle and employed when the solution is used in sugar determination.

Sample Preparation

Weigh 25 g of juice, jam, marmalade, fruit ... and transfer to 250 ml volumetric flask using about 100 ml water. Blend the mixture and neutralize with 1 N NaOH. Make up to volume and filter through No. 4 Whatman paper. Pipette a 100 ml aliquot into a 500 ml volumetric flask. Add 2 ml of neutral lead acetate solution and about 200 ml of water. Allow to stand for 10 min, then precipitate the excess lead with potassium oxalate solution. Make up to mark and filter.

Standardization of Soxhlet-Reagent

Accurately pipette 10 ml mixed Soxhlet-reagent (5 ml each of Fehling's solutions) into a 250 ml Erlenmeyer-flask. Place the standard invert sugar solution in a 50 ml burette. Add to the Soxhlet-reagent almost the whole of the standard invert sugar solution (18 to 19 ml) required to effect the reduction of all the copper, so that not more than 1 ml will be required later to complete the titration.

Heat the flask containing the cold mixture over a burner covered with asbestos filled wire gauze. When the liquid begins to boil, keep it in moderate ebullition for 2 min without removing from the flame, and add 3 drops of methylene blue indicator.

Complete the titration in a further minute, so that the reaction mixture boils altogether for 3 min without interruption.

The end point is indicated by the decolourization of the indicator. Note the volume of the sugar solution required for reducing 10 ml of Soxhlet-reagent completely. The equivalent volume should be 20.37 ± 0.05 ml. Small deviations may arise from variations in the individual procedures or composition of the reagents.

$$a = b \, 0.0025$$

where:

a = Factor for Soxhlet-reagent (g of invert sugar)
b = Titration

Determination of reducing sugar in the sample: The sugar solution should be neutral and its concentration should be such that the titre value ranges between 15 and 50 ml. For this purpose, adjust the sugar concentration in the solution taken for titration so that it contains 0.1 to 0.3 g of sugar per 100 ml, when 10 ml of mixed Soxhlet-reagent is used.

If the approximate concentration of sugar in sample is unknown, proceed by the *incremental method* of titration. When the correct dilutions are established perform subsequent titrations by the standard method of titration.

The Incremental Method: Pipette 10 ml of the mixed Soxhlet-reagent into a 250 ml flask and add 50 ml water. Fill the burette with the clarified sugar solution. Add from the burette 15 ml, mix and heat to boiling. Boil for 15 s. If the colour remains blue (indicating that the Soxhlet-reagent is not completely reduced), add further 2–3 ml of the sugar solution. Boil the solution for a few seconds until only a faintest perceptible blue colour remains. Add 3 drops of methylene blue solution and complete the titration by adding the sugar solution drop wise until the indicator is completely decolourized.

Record the volume of solution required. The accuracy of the incremental method is increased by attaining the end point as rapidly as possible and by maintaining a total boiling period of 3 min. Error resulting from this titration will generally be 1%.

The Standard Method: Pipette 10 ml of mixed Soxhlet-reagent into each of two 250 ml conical flask. Fill the burette with the solution to be titrated. Run into the flask almost the whole volume of sugar solution required to reduce the Soxhlet-reagent, so that about 1 ml is required later to complete the titration. Mix the contents of the flask, heat to boiling and boil moderately for 2 min. Then add 3 drops of the methylene blue solution.

Complete the titration within 1 min by adding 2 to 3 drops of sugar solution at 5 to 10 s intervals, until the indicator is completely decolourized.

At the end point, the boiling liquid assumes the brick-red colour of precipitated cuprous oxide. Note the volume of the solution required.

$$a = \frac{b\ 100}{c}$$

where:

a = mg sugar/100 ml
b = Total reducing sugar required
c = Titration

2.3.2 Determination of Total Sugars

Pipette 50 ml of the clarified solution into a 250 ml conical flask. Add 5 g of citric acid and 50 ml of water. Boil gently for 10 min to complete the invertion of sucrose, then cool. Transfer to a 250 ml volumetric flask and neutralize with 1 N NaOH using phenolphthalein as indicator. Make up to volume and take an aliquot to determine the total sugar as invert sugars.

Calculation:
$$a = \frac{b\ c}{d\ e}$$

where:

a = % Reducing sugar
b = mg of Invert sugar
c = Dilution
d = Titration
c = Wt of the sample

2.3.3 Chromatographic Analysis of Sugars

Occasionally it necessary to separate, identify and determine quantitatively the individual sugars in a fruit or a fruit product. Sugars can be separated and detected quantitatively by both Gas Liquid Chromatography (GLC) and High Performance Liquid Chromatography (HPLC).

The GLC-Analysis of trimethylsilyl-(TMS)-ethers of sugars with Trisil as the silylating agent and myoinositol as the internal standard is frequently used [9].

Add 2 ml of Trisil to the sugar extract/myoinositol mixture and allow to react for 30 min at 45 C. Inject typically 0.5–1.0 µl of the silylated mixture into a gas chromatograph with a flame ionization detector. The carrier gas (nitrogen) flow rate is 25 cm^3/min and hydrogen gas flow rate 45 cm^3/min. A stainless steel column (1 m x 2.4 mm i.d.) packed with a suitable material such as:

– 3.8% SE-30 on chromosorb W HDMS 60/80
– 3.0% OV-17 on chromosorb G A/W HDMS 60/80

The column can be operated isothermally at 170 C and 225 C with detector temperature at 270 C and injector temperature at 250 C.

HPLC-Analysis of Sugars is today the recommended method. The advantages of HPLC over GLC of trimethylsilylether of sugars include faster preparation, ability to chromatogram aqueous solutions, and the direct, non-derivative analytical route. Furthermore, HPLC-chromatograms are more easily interpreted since each sugar yields only one peak, whereas anomerization may result in as many as seven TMS peaks per monosaccharide [10].

Preparation of the sample requires only aqueous extraction at 65 C, centrifugation (10 min at 48 000 rpm), filtration and dilution with acetonitrile. The volume of water used is such that the concentration of sugar is within 1–5% range. Injection of 15 µl of sample yield readily measurable peak height.

The recommended solvents are acetonitrile/water (3:1 v/v or 4:1 v/v) in a flow rate of 2 ml/min, and columns are Spherisorb S$_5$ NH$_2$ (25 cm x 4.6 mm i.d.) or BONDAPAK/Carbohydrate (Waters Associates, Milford, MA, USA). The detection of sugar is most commonly carried out by use of refractive index monitoring. Ultraviolet detection is probably not practicable for routine use, since a detection wavelength below 200 nm must be used, which means that solvent purity is exceptionally critical. Generally a differential refractometry (e.g. Waters Associates R 401) having a sensitivity of 1 x 10^7 refractive units is used.

Figure 3 shows a typical separation of fructose, glucose, sucrose, maltose and lactose standards [10].

The number of publications on the use of HPLC for sugar determination has increased significantly over the past few years. It may be significant to note that the most of these articles appeared in the Journal of the Association of Official Analytical Chemists (JAOAC) and in Journal of Chromatography. The following publications can be recommended: Hurst and Martin [11], DeVries et al. [12], Binder [13], Damon and Pettitt [14].

2.3.4 Enzymatic Analysis of Sugars

The large number of methods that are nowadays used routinely and in research proves that enzymatic analysis has brought a genuine enrichment of fruit juice analysis, since it has made it possible to obtain results rapidly and specifically with very little expenditure on personnel and apparatus [15]. Sucrose, D-glucose and D-fructose can be determined enzymatically using single reagents [16].

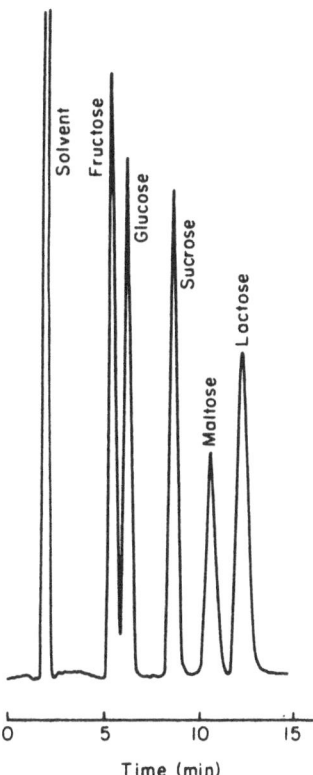

Fig. 3. Analysis of common food sugars. Column, Spherisorb S$_5$ NH$_2$, 250 mm x 4.6 mm; solvent, 75 % acetonitrile, 25 % water /v/v), 2 ml min^{-1}; refractive index detection [10]

Principle [17]: The D-glucose content is determined before and after the enzymatic hydrolysis of sucrose; D-fructose is determined subsequently to the determination of D-glucose. Determination of D-glucose before inversion: At pH 7.6 the enzyme hexokinase (HK) catalyzes the phosphorylation of D-glucose by adenosine-5'-triphosphate (ATP) with the simultaneous formation of adenosine-5'-diphosphate (ADP). In the presence of glucose-6-phosphate dehydrogenase (G6P-DH) the glucose-6-phosphate (G6P) formed is specifically oxidized by nicotinamide-adenine dinucleotide phosphate (NADP) to gluconate-6-phosphate with the formation of reduced nicotinamide-adenine dinucleotide phosphate (NADPH). The NADPH formed in this reaction is stoichiometric with the amount of D-glucose and is measured by means of its absorbance at 334, 340 or 365 nm.

Determination of D-fructose: Hexokinase also catalyzes the phosphorylation of D-fructose to fructose-6-phosphate (F6P) with the aid of ATP. On completion of the last to reactions F6P is converted by phosphoglucose isomerase (PGI) to G6P. G6P reacts again with NADP under formation of gluconate-6-phosphate and NADPH. The amount of formed now is stoichiometric with the amount of D-fructose.

Enzymatic inversion: At pH 4.6 sucrose is hydrolysed by the enzyme invertase (fructosidase) to D-glucose and D-fructose. The determination of D-glucose after inversion (total D-glucose) is carried out according to the principle outlined above. The sucrose content is calculated from the difference of the D-glucose concentration before and after inversion.

Reagents

- Citric acid.
- Trisodium citrate
- Sodium hydroxide, 2.0 mol/l
- Magnesium sulphate
- Sodium hydroxide, 5 mol/l
- Sodium hydrogencarbonate
- Sucrose, D-glucose and D-fructose for standard solution.
- Invertase (fructosidase)
- Triethanolamine hydrochloride
- NADP-NA$_2$
- ATP-NA$_2$H$_2$
- HK/G6P-DH
- PGI

Preparation of solutions (for ca. 50 Determinations):

- Citrate buffer; 0.32 mol/l, pH 4.6: Dissolve 6.9 g citric acid and 9.1 g trisodium citrate with ca. 150 ml redist. water, adjust to pH 4.6 with sodium hydroxide (2.0 mol/l), and fill up to 200 ml with redist. water.
- Invertase solution; 2.5 mg/ml, 750 U/ml: Dissolve 5 mg lyophiliste with 2 ml citrate buffer. The invertase solution is stable for 1 week at 4 C.
- Triethanolamine buffer (Tra buffer); 0.75 mol/l, pH 7.6: Dissolve 14 g triethanolamine hydrochloride and 0.25 g Magnesium sulphate with 80 ml redist. water, adjust to pH 7.6 with ca 5 ml Sodium hydroxide solution (5 mol/l), and fill up to 100 ml with redist. water. The buffer is stable for four weeks at 4 C. Bring buffer solution to 20–25 C before use.
- NADP solution; ca. 11.5 mmol/l: Dissolve 60 mg NADP-Ha$_2$ with 6 ml redist. water. The solution is stable for four weeks at 4 C.
- ATP solution, ca. 81 mmol/l: Dissolve 300 mg ATP and 300 mg Sodium hydrogen carbonate in 6 ml redist. water. The solution is stable for four weeks at 4 C.
- HK/G6P-DH solution: 2 mg HK/ml; 1 mg G6P-DH/ml. Use the suspension undiluted. This suspension is stable for one year at 4 C.
- PGI solution: 2 mg PGI/ml. Use the suspension undiluted. The suspension is stable for one year at 4 C.

– Standard solutions (prepare freshly before use): Measurement of the standards is only for checking the working technique.
– 1 mg sucrose/l: Dissolve 100 mg sucrose in 100 ml water.
– 1 g D-glucose/l: Dissolve 100 mg D-glucose in 100 ml water.
– 1 g D-fructose/l: Dissolve 100 mg D-fructose in 100 ml water.

Procedure: Filter turbid juices (alternatively clarify with Carrez reagents- s. below) and dilute sufficiently to yield a sucrose, D-glucose and D-fructose concentration of approx. 0.1–1.5 g/l. The diluted sample solution can also be used for the assay if it is coloured. Only strongly coloured juices which are used undiluted for the assay because of their low sugar content are to be decolorized as follow: Mix 10 ml of juice and ca. 0.1 g of polyamide powder or polyvinyl polypyrrolidone, stir for 1 min and filter. Measurement is made at the absorption maximum of 340 nm when spectrophotometers are used and at 365 or 334 nm when a spectral-line photometer with an Hg lamp is used. Use a glass cuvette of 1 cm light path at 20–25 C. The final volume should be 3.14 ml and read against air without a cuvette in the light path or against water.

For further information read the publications of Boehringer [16], Bergmeyer [17], and ASU [18].

2.4 Starch Test

Most of the tropical fruits, especially early in the season, contain starch. This gives rise to viscosity and sedimentation (fouling) on the surface of pasteurizers and evaporators, mainly due to the gelatinization of starch. This means a reduction in quality, a reduction in heat transfer and cleaning the equipment at frequent intervals.

Up to 95% of the starch of fruit juices can be eliminated by centrifugation before processing (before gelatinization). Commercial enzymes, mainly amylase, can be used for the breakdown of the starch. It can be also done simultaneously with the breakdown of pectin (NOVO Ferment and Röhm).

Iodine – Test for Starch

Heat about 10 ml of the juice up to 70 C, except for processed juices, cool and mix with a few drops of iodine solution (aqueous solution of 1% iodine and 10% potassium iodide).

Yellow colouring: the juice contains no more starch but some dextrine.

Brown or blue colouring: the starch has not been completely broken down.

2.5 Determination of Alcohol Insoluble Solids

The quality of fresh fruits is determined by the maturity at the time of harvest. The maturity is related to the alcohol-insoluble solids (AIS) which consist of starch, hemicellulose, cellulose and pectin. The ripe fruits have lower AIS-values than mature ones.

The amount of alcohol-insoluble solids and also the water-insoluble solids is probably the most commonly used factor for the calculation of the amount of fruit used in the manufacture of jams and jellies. The results obtained by this procedure are sometimes called "Crude pectin".

Insoluble solids, which consist of pulp, affect the appearance and "mouth feel" of most tropical fruit juices. They also effect the consistency.

The determination of suspended pulp content [7] of a juice may carried out by the use of a centrifuge. Fill the centrifuge tubes with 50 ml juice and centrifuge for ten minutes. The speed is adjusted to 1600 rpm (revolution per min.) for a 10-inch (about 25 cm) diameter centrifuge (distance between the bottoms of opposing centrifuging tubes in the horizontal operating position). The speeds with other centrifuge sizes are 1313 rpm for 15 inches (38 cm) and 1137 rpm for 20 inches (50 cm).

According to the U.S. Standard, pineapple juice should contain not less than 5% and not more than 30% of finely divided insoluble solids.

Determination of Alcohol Insoluble Solids [19]

Weigh 20 g of ground fruit material in a 0.5 l beaker. Add 300 ml of 80% alcohol, stir, cover the beaker and bring to boil. Simmer slowly for 30 min. Fit into a Buchner funnel, a filter paper which has been previously dried and weighed. Apply suction and transfer the contents of the beaker. Wash the residue on the filter paper with 80% alcohol until the alcohol is clear and colourless when it has passed through the filter paper. Dry the filter paper for 2 h at 100 C.

A useful method often employed involves extracting the sample in a dried, weighed Soxhlet thimble with 80% alcohol (or water for the determination of water-insoluble solids). The insoluble residue is treated with acetone before drying in an oven at 100 C.

2.6 Determination of Acidity

2.6.1 Determination of Titratable Acidity

Titratable acidity may be expressed as the amout of free acid (mainly as anhydrous citric acid) in the product (g/100 g, g/100 ml, or g/l).

Equipment:

– Automatic burette and reservoir
– pH-Meter
– Analytical balance
– 10 ml Transfer pipette
– 250 ml Erlenmeyer flask

Reagents:

– Sodium hydroxide solution 0.1 N (4 g NaOH/l)
– Phenolphthalein indicator, 1% solution in ethanol, neutralized with sodium hydroxide to faint pink.

Procedure: Pipette 10 ml juice into a 250 ml Erlenmeyer flask, add 50 ml water and 5 drops of indicator. 10 g of fresh or dried fruits or marmalades can be mixed with hot water in a blender, cooked and filtered if necessary.

Titrate against 0.1 N NaOH quite rapidly until near pH 6, then add the alkali slowly to pH 7. Finish titration by adding 4 drops at time until pH 8.1. If no pH-meter available, titrate to the desired end point (faint pink colour).

Calculation: If using 10 ml of juice and 0.1 N NaOH, multiply ml of NaOH used by 0.064 to obtain gm of citric acid/100 ml (0.075 for tartaric acid and 0.09 for lactic acid)

$$a = \frac{b\,c\,d\,e\;100}{f}$$

where:

a = % Acidity
b = Titration
c = Normality of alkali
d = Volume made up
e = Equivalent wt of acid
f = Wt or volume of sample taken

Some standard methods have the end point of the titration at pH 7.0 [20]. Therefore it is especially important to compare only results obtained using the same conditions. When the end point of phenolphthalein indicator is reached the pH meter should read 8.2.

2.6.2 Determination of Organic Acids

Sometimes it is necessary to separate, identify and determine quantitatively the individual fruit product. Fruit juice concentrates may contain citric acid added as an acidulant, or to conceal watering of the juice. Citric acid is dominant in citrus fruits, malic acid in apples, and isocitric acid in blackberries. Adulteration of fruit juices can be detected by determining the relation between the organic acids, e.g. Citric/Malic or Citric/Isocitric. In such cases, comparison must be made with analysis of authentic juices, and methods must be available for the determination of small quantities of some organic acids in the presence of larger quantities of other organic acids. Organic acids can be separated and detected quantitatively by Gas Liquid Chromatography (GLC), High Performance Liquid Chromatography (HPLC), and by Enzymatic Methods. In contrast to the traditional chemical and physico-chemical analytical techniques, the use of GLC, HPLC and Enzymatic Analysis is becoming more and more widespread in fruit processing plants. The main reason for this is that the technique is fast, usually taking no more than 30 minutes per analysis and that it provides information which cannot easily be obtained in any other way [21].

2.6.2.1 Gas-Liquid Chromatography (GLC)

Extraction of Organic Acids

Add 10 ml of absolute alcohol to a 20 g sample in a micro-blender jar, following by adding 50 ml of 80% alcohol and blending. Filter and wash the filter with alcohol. Transfer the combined filtrate to a 100 ml volumetric flask and make up to volume with 80% alcohol.

Add five grams of Dowex MSC-1 cation exchange resin (in H-form) to the filtered extract in a 250-ml beaker. Mix with a magnetic stirrer and then filter. Rinse three times with water. Pass the combined filtrate through an anion exchange resin, acetate form Amberlite IRN-78 packed in a 10 cm x 1 cm column, and allow to flow at a rate of approximately 1 ml/min. Wash with water. Eluate the organic acid from the resin with 100 ml of 6 N formic acid following by water at the same flow rate. Dry the eluate under vacuum in a rotary evaporator at 50 C. To insure removal of moisture and formic acid, store the sample over night in a vacuum oven with a mixture of sodium hydroxide and calcium chloride (1:2) as the desiccant.

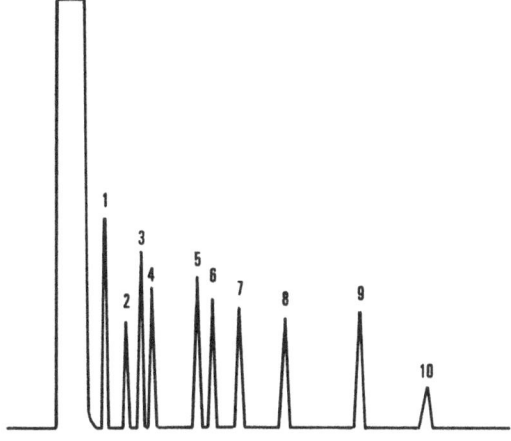

Fig. 4. GLC chromatogram of standard mixture of organic acids as TMS derivatives. Peak 1) lactic, 2) oxalic, 3) naphtalene as internal standard 4) malonic, 5) succinic, 6) fumaric, 7) glutaric, 8) malic, 9) tartaric and 10) citric [23]

Preparation of Trimethylsilyl (TMS) Derivatives

Add 5 g of methoxyamine hydrochloride to the flask containing the dried purified extract. Add 1 ml of naphthaline as internal standard (5 mg/ml in dried pyridine). To dissolve the residue allow the mixture to stand for 10 min at room temperature. Add 1 ml of *N,O*-Bis-(Trimethylsilyl)-acetamide. Heat the solution in a stoppered flask in an oil bath for 30 min at 70 C. Cool and use the solution for the GLC analysis [22].

Gas-Liquid Chromatography

Column: U-shaped stainless steel column (1.8 m x 0.64 cm o.d.) packed with 3% SE-52 coated on high performance, acid washed DMCS treated Chromasorb W (Hewlett-Packard).

Temperature: from 90 to 190 C at the rate of 2 C/min with the injection port and the flame ionization detector at 200 and 270 C, respectively.

Flow rates: 55 ml/min for the nitrogen carrier gas, 35 ml/min for the hydrogen and 500 ml/min for the air.

A typical chromatogram of the standard solution of the organic acid is presented in Fig. 4. The GLC method is described by Mabesa et al. [23].

2.6.2.2 High Performance Liquid Chromatography (HPLC)

HPLC is often the technique of choice for organic acid determination in fruit and fruit products.

Sample clean-up Sample clean-up required for fruit juice analysis is minimal. In some cases, it is possible to dilute the juice to ten-fold with water, filter, and inject straight into the chromatograph. For juices high in polyphenols, however it is probably worth removing them by pretreatment on a polyamide column. This will help to prolong the life of the column. For the elimination of sugars and free amino acids the extraction method described in Sect. 2.6.2.1 is recommended.

HPLC Separation

The two main column types for organic acid separations are Reverse-Phase-C8 or -C18 Hydrocarbons bonded to a silica core, or Cation-Exchange-Resins operated in the hydrogen form. These resins are similar to those used for sugar analysis and are offered pre-packed by the same suppliers [24]. Examples of separation obtained on both types of column are shown in Figs. 5 and 6. The reverse-phase columns are operated in aqueous buffers at between pH 2 and 3, typically with 0.01 M to 0.02 M potassium hydrogen phosphate adjusted to the correct pH with phosphoric acid. In some cases the addition of tetrabutyl ammonium phosphate (5 mM) is recommended, to form ion-pairs with the acids being chromatographed and therefore to increase their retention times and improve separation [10, 24, 25]. As usual with such columns, the optimum operating temperature is around 45 C. Figure 5 shows the separation of organic acids on a Spherisorb ODS-2-column in 0.01 M potassium hydrogen phosphate, pH 3, with 0.001 M tetrabutyl-ammonium phosphate [24].

Using a similar phosphate buffer and a LiChrosorb RP8 column (250 mm x 4.6 mm i.d.) Jeuring et al. [25] have separated malic and citric acid in apple juice.

Employing Ion-exchange resins for the separation organic acid is also usual. Figure 6. shows the analysis of organic acids on a Aminex HPX-87 column, with .0065 M sulphuric acids as eluent at 65 C, flow rate 0.8 ml/min and with UV-detection at 210 nm [26].

UV detectors are usually used for the detection and determination of organic acids between 190 and 230 nm. They can be successfully operated at 220–230 nm to detect organic acids without removal of sugars, but it may be necessary to remove UV absorbing amino acids. If greater sensitivity is required, the detector may be tuned to 210 nm but at this wavelength sugars such as fructose have appreciable absorption and may interfere. Therefore, clean-up will be almost always required to remove sugars and amino acids. More about HPLC techniques is given by Engelhardt [27].

2.6.2.3 Enzymatic Determination

Todays several organic acids can be individually determined by enzymatic methods, e.g. acetic, citric, formic, lactic, malic, pyruvic and succinic acids [15, 16]. For a tropical fruit processing plant the determination of citric acid is the

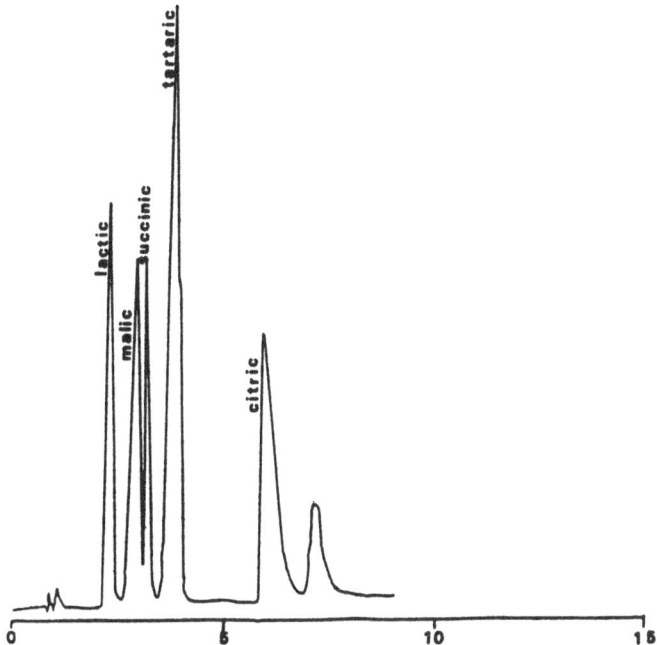

Fig. 5. Separation of organic acids on a Spherisorb ODS-2-Column in 0.01 M potassium hydrogen phosphate, pH 3, with 0.001 M tetrabutyl ammonium phosphate [24].

most important one. This method is officially recommended in Germany [28], and is specific for citric acid.

Principle [29]: Citric acid (citrate) is converted to oxaloacetate and acetate in the reaction catalyzed by the enzyme citrate lyase (CL). In the presence of the enzyme malate dehydrogenase (MDH) and L-lactate dehydrogenase (L-LDH),

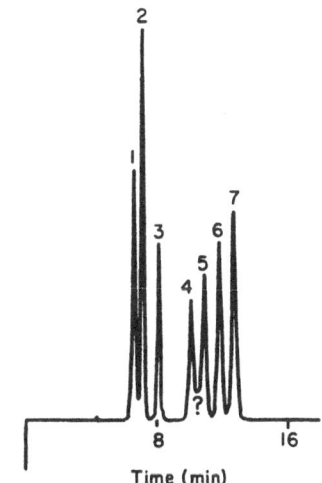

Fig. 6. Separation of organic acids [1 citric, 2) tartaric, 3) malic, 4) succinic, 5) lactic, 6) fumaric, 7) acetic] on a Aminex HPX-87 column (300 mm x 7.8 mm) using 0.0065 M sulphuric acid as eluent

oxaloacetate and its decarboxylation product pyrovate are reduced to L-malate and L-lactate, respectively, by reduced nicotinamide-adenine dinucleotide (NADH). The amount of NADH oxidized is stoichiometric with the amount of citrate. NADH is determined by means of its absorbance at 334, 340 or 365 nm.

Reagents:

- Sodium hydroxide, 5 mol/l
- Zinc chloride
- Sodium hydrogen carbonate
- Citric acid monohydrate for the preparation of standard solution
- Glycylglycine
- NADH-Na$_2$
- MDH
- LDH
- CL

Preparation of solutions (for ca. 50 determinations):

- Buffer (Glycylglycine, 0.54 mol/l, pH 7.8, Zinc chloride, 0.6 mmol/l): Dissolve 7.13 g glycylglycine in ca. 7 ml water, adjust to pH 7.8 with ca. 13 ml NaOH (5 mol/l), add 10 ml zinc chloride solution (80 mg zinc chloride/100 ml), and make up to 100 ml with water. The buffer is stable for four weeks at 4 C. Bring buffer solution to 25 C before use.
- NADH solution (ca. 6 mmol/l): Dissolve 30 mg ADH-Na$_2$ and 60 mg sodium hydrogen carbonate in 6 ml water. The solution is stable for four weeks at 4 C.
- MDH/LDH solution (1 mg MDH/ml; 4 mg LDH/ml): Mix 0.1 ml MDH (5 mg/ml and 0.4 ml LDH (5 mg/ml). The suspension is stable for one year at 4 C.
- CL solution (10 mg protein/ml = 320 mg lyophilisate/ml = 80 U/ml): Dissolve 160 mg of lyophilisate (= 5 mg enzyme protein) in 0.5 ml of ice-cold water. The solution is stable for one week at 4 C, and four weeks at –20 C.
- Standard solution (0.4 g citric acid/l): Dissolve 43.4 mg citric acid monohydrate in 100 ml water.

Procedure: Measurement is made at an absorption maximum of 340 nm when spectrophotometers are used, and at 365 or 334 nm when a spectral-line photometer with an Hg lamp is used. Use a glass cuvette of 1 cm light path at 20–25 C. Read against air (without a cuvette in the light path) or against water.

Remove turbidities by filtration and dilute the juice sample to obtain a citric acid concentration between 0.02 and 0.4 g/l. The diluted solution can be used for the assay even if it is coloured. Only intensively coloured juices must be decolourized when they are used undiluted for the assay because of their low citric acid concentration. In such cases, proceed as follows: Mix 10 ml juice and 0.1 g polyamide or polyvinyl polypyrolidone (PVPP), stir for 1 min, and filter.

Use the clear, slightly coloured solution for the assay, neutralize, if necessary. The Carrez-clarification cannot be used in the sample preparation for citric acid determination due to a low recovery rate (adsorption of citric acid).

More about the enzymatic determination of citric acid is given by ASU [28], Mollering [29] and Boehringer [16].

2.7 The Brix/Acid Ratio

The Brix/acid ratio is used as an index of maturity by the tropical fruit processing industry. To the buyers of tropical fruit products it is an indication of the relative sweetness or tartness of a product. The higher the Brix in relation to the acid content of the juice, the higher the ratio and the "sweeter" the taste.

To obtain Brix/acid ratio, the percent of soluble solids or Brix is divided by the percent of titrable acidity:

$$a = \frac{-8 \quad b}{c}$$

where:

a = Brix/acid ratio
b = Brix value
c = Acidity g/100g

Example: for mango pulp the Brix degree ist 12.0, the acidity is 0.6 g/100 g, and the Brix/acid ratio is 20.0:1. For orange juice the ratio is usually 14.5:1.

2.8 Measurement of pH

The term pH is the symbol for hydrogen-ion concentration, and is defined as the logarithm of the reciprocal of hydrogen ion concentration in g per liter. The pH scale is logarithmic, therefore, pH 5.0 is 10 times as acid as pH 6.0.

pH is of importance as a measure of the active acidity which influences the flavour or palatability of a product and affects the processing requirements. The most important factor affecting the sterilization time and temperature and the use of preservatives, is the actual pH-value in fruit products. It is usually considered that a pH of 4.6 is the dividing line between acid and nonacid foods.

Adjustment of the pH-value is necessary for the production of jams, jellies, marmalades and soft drinks.

There are two methods used to measure the pH of Food. One is the colourimetric method using an indicator solution or paper. The other method is to measure the potential developed between two electrodes, when immersed in a solution. The glass electrode pH-Meter has largely replaced the older methods, and is the quickest and most reliable method available. There are a number of pH-Meters on the market, from portable to line operated models. Modern pH-Meters have digital readout systems. If the mains voltage is likely to be unstable the pH-Meter should be fitted with a voltage regulator. The batteries in battery operated pH-Meters should be checked at regular intervals. Calomel electrodes are often used as the reference electrode and they should be kept filled with saturated potassium chloride solution or a solution specified by the manufacturer. The electrodes should be soaked in buffer solution, water or other liquid specified by the manufacturer. Buffers for the standardization of the pH-Meters are available on the market, or they can be freshly prepared.

Buffer Solutions for Standardization of the pH-Meter

Usually, two solutions will suffice to standardize high and low pH-values. The majority of the buffer solutions are not stable for a long time and should be freshly prepared. Addition of a drop of pure toluene will inhibit mould growth. Many solutions, particularly those above pH 7, have a fairly large pH/temperature coefficient. It is, therefore, important to use the pH value of the standardizing buffer at the same temperature as that of the sample (usually 20 C), and to state the temperature in quoting the results.

Preparation of Buffer Solutions

0.05 M Potassium acid phthalate solution-pH 4 at 20 C: Weigh accurately 10.207 g of potassium acid phthalate, $KHC_8H_4O_4$, AR grade, and dissolve in freshly boiled and cooled water, make up the volume to one liter and add one drop of toluene and shake well.

0.05 M Sodium borate solution- pH 9.18 at 20 C: Dissolve 19.017 g of sodium borate, $Na_2B_4O_7 . 10 H_2O$, Ar grade, in water and make up the volume to 1 litre. The water must be freshly boiled (CO_2-free). Add a drop of toluene and shake well.

Saturated potassium chloride solution for calomel electrode: Pure crystals of KCl of AR grade in water. Always keep 1 cm layer of crystals in the bottle.

2.9 Determination of Ash

The ash is the inorganic residue remaining after the organic matter has been burnt away. The ash content, water soluble ash and alkalinity of the ash can be regarded as a general measure of the quality and purity of fruit juices, fruit juice powders, sugars, citric acid and other food additives. The determination of acid insoluble ash is also a good measure of quality, and of the sand and other matter present.

Equipment:

- Muffle Furnace (800 C)
- Crucibles (30 ml) or silica dishes (8–10 cm diameter)
- Tongs
- Steam bath or water bath
- Analytical balance
- Desiccator

Preparation of Crucibles: Heat crucibles to 800 C for 2–3 hours. Allow crucibles and muffle furnace to cool to 200 C. Using tongs, remove crucibles from muffle and place in desiccator until they have reached room temperature.

Determination of Ash: Weigh 10 g of the sample and place in the crucible on a steam or water bath until the sample is dry. Transfer the crucible to muffle furnace and raise temperature slowly to avoid spatting of sample, or heat the crucible on a bunsen burner and then in the muffle furnace. Maintain temperature at 525 C (max. 550 C) for 6 hours. Ashed sample will be white to slightly grey when ashing is complete. If ashing is not complete, break up the ash with a platinum wire, and incinerate once more. Cool crucible in desiccator to room temperature and reweigh it.

Water-soluble ash and water-insoluble ash: The ash is boiled with 25 ml water, filter through an ashless filter paper and wash with 50 ml hot water. Transfer the filter paper to a crucible, incinerate, cool and weigh the water-insoluble ash. Calculate the water-soluble ash.

Acid-insoluble ash: Add 25 ml of dilute HCl (10% w/w HCl) to the ash and boil gently for 5 min. Filter through an ashless filter paper and wash with hot water. Return the paper to the crucible, incinerate, cool and weigh the insoluble ash.

Alkalinity of the ash: The alkalinity of the total ash or of the insoluble ash can be obtained by titrating with 0.1 N sulphuric acid or 0.1 N HCl using methyl orange as indicator.

The alkalinity can be expressed as the number of ml of 0.1 N acid required to neutralize the ash from 100 g of the sample, or as alkalinity number (alkali number), number of ml 1 N acid required to neutralize 1 g ash.

The recommended method for the determination of alkalinity of the ash is to add 10 ml of 0.1 N HCl or H_2SO_4 to the ash Dissolve by warming on a water bath, cool and titrate the excess acid with 0.1 N NaOH using methyl orange as indicator (the alkali consumed = X). Carry out a blank titration using 10 ml of 0.1 N HCl (the alkali consumed = Y). The alkalinity can be calculated as Potassium carbonate, K_2CO_3, during ashing citrates are transformed to carbonates.

One ml of 0.1 N acid is equal to 0.00691 g of K_2CO_3. The alkalinity can be expressed as g of Potassium carbonate per g of ash

$$a = \frac{(Y-X)\,0.00691}{b}$$

where:

a = Alkalinity of ash
Y = The alkali consumed in blank titration
X = The alkali consumed
b = Weight of ash

2.10 Determination of Vitamin C

Tropical fruits are important sources of Vitamin C. The Vitamin C activity of foods may be derived from both L-ascorbic acid and its oxidation product L-dehydroascorbic acid. *Dehydroascorbic acid* is unstable and readily converts to diketogulonic acid which does not possess Vitamin C activity. Dehydroascorbic acid is easily reduced back to ascorbic acid by reducing agents such as hydrogen sulphide or homocysteine.

D-Ascorbic acid (iso-ascorbic acid, erythorbic acid) has a biological activity which is about 20 times less than that of L-ascorbic acid. It does not occur in natural products but is sometimes added to fruit products as an antioxidant.

Vitamin C is sensitive to heat, oxygen and light. Therefore, its determination during processing and storage is an index for the quality and a measure of possible deterioration of most fruit products.

The ascorbic acid content can be estimated by macerating the sample with a stabilizing agent such as metaphosphoric acid (to maintain proper acidity for reaction and to avoid autooxidation of ascorbic acid at high pH) and reduction of 2,6-dichlorophenol-indophenol. This dye which is blue in alkaline solution and red in acid solution, is reduced by ascorbic acid to a colourless form. The reaction is practically specific for ascorbic acid in solutions in the pH range 1–3.5 ([8, 30].

Reagents:

- Metaphosphoric acid (HPO_3) solution, 3%
- *Ascorbic acid standard:* Dissolve 100 mg of L-ascorbic acid in 100 ml of 3% HPO_3. Dilute 10 ml to 100 ml with 3% HPO_3 immediatley before use (1 ml = 0.1 mg of ascorbic acid).
- *Dye solution:* Dissolve 50 mg of sodium salt of 2,6-dichloro-phenol-indophenol in 150 ml hot water containing 42 mg sodium bicarbonate. Cool and dilute with water to 200 ml. Store in a refrigerator and standardize every day before use. To standardize the dye, take 5 ml of standard ascorbic acid solution and add 5 ml of HPO_3. Fill a microburette with the dye, titrate with the dye solution to a pink colour which should persist 15 seconds. Express the concentration as mg ascorbic acid equivalent to 1 ml of the dye solution.

$$a = \frac{0.5}{b}$$

where:

    a = Dye factor
    b = Titration

Determination: Take 10 to 25 ml of the juice and make up to 100 ml with 3% HPO_3, filter or centrifuge. For dried fruits blend the sample with HPO_3. Take an aliquot (5 ml) of the HPO3 extract of the sample, add 2.5 ml of acetone and titrate with dye until a faint pink colour persists for 15 seconds. Calculate the vitamin C content as mg per 100 ml (or 100 g). The acetone may be omitted if sulphur dioxide is known to be absent. Its function is to form the acetone bisulphite complex with sulphur dioxide which otherwise interferes with the titration. The same can be done by formaldehyde condensation, adding 1 ml of 40% formaldehyde and 0.1 ml of HCl, keeping for 10 min and titrate as before.

Calculation:

$$a = \frac{b\,c\,d\,100}{e\,f}$$

where:

    a = Ascorbic acid mg/100 g or ml
    b = Titration
    c = Dye factor
    d = Volume made up
    e = Aliquot of extract
    f = Wt or Volume of sample

Determination of Dehydroascorbic acid

Estimate the ascorbic acid as already described, then through proportion of the solution pass a stream of hydrogen sulphide for 10 min. Stopper the flask and allow to stand overnight in a refrigerator, remove the hydrogen sulphide by bubbling nitrogen through the mixture and titrate as before. The difference between the two titrations gives a measure of the dehydroascorbic acid.

2.11. Determination of Carotenoids

The main natural colours in food are chlorophylls, carotenoids, anthocyanins and riboflavins. The determination of total carotenoids, and possibly chlorophylls, as well in tropical fruits and their products is a measure of their stage of maturity and of the quality. Carotenoids are sensitive to oxidation and are readily bleached when exposed to oxygen and sunlight. However, the use of antioxidants such as ascorbic acid and tocopherols can reduce this oxidative degradation substantially. Chlorophylls and their copper-complexed derivatives, exhibit fairly good heat and light stability, but their stability toward acid are poor. Chlorophylls are easily degraded in acid solution to olive-brown compounds, known as pheophytins [8, 30, 31].

Determination of total Carotenoids and Chlorophylls

Ten grams of the sample are mixed with 30 ml of 85% acetone in a dark bottle and left to stand for 15 hours at room temperature. The sample is then filtered through glass wool into a 100 ml volumetric flask, and made up to volume with 85% acetone solution. The crude pigment is assayed spectrophotometrically using the following equations:

After measurements at 440, 644 and 662 nm:

Chlorophyll a = $(9.784 \times E_{662}) - (0.99 \times E_{644})$ = mg/liter
Chlorophyll b = $(21.426 \times E_{644}) - (4.65 \times E_{662})$ = mg/liter
Carotenoids = $(4.695 \times E_{440}) - 0.268$ (Chl. a + Chl. b) = mg/liter

where:

E = sample optical density at the indicated wave length.

For more accurate determination the acetone extract has to be concentrated in a bench-type rotating vacuum evaporator. Fifty ml of freshly distilled n-hexane (69–70 C) is added. The extract is washed twice with 20 ml portions of methanolic potassium hydroxide solution (100 g potassium hydroxide dissolved in 750 ml methyl alcohol and 250 ml water). The aqueous layer is introduced into a separatory funnel and the solution is dried over sodium sulphate.

The purification of the crude pigments can be done by adsorption on to magnesia instead of saponification with methanolic potassium hydroxide. The concentrated crude pigment is dissolved in 15 ml hexane and is poured on a 2 x 20 cm column of Florisil (60 to 80 mesh, activated at 520 C) and washed with hexane. Washing with hexane removes only a small portion of the total colour and is continued until the effluent is nearly colourless. The column is subsequently washed with 10:90 acetone:hexane and then with 30:70 acetone-hexane to remove the adsorbed carotenoids. The eluted fractions are concentrated under vacuum to remove the solvent.

The extraction of the carotenoids directly from fruit juices is also possible by adsorption on the magnesia (Fisher S-120 infusorial earth), the fruit juice is drawn through a layer of magnesia which is then extracted with dichloromethane-methanol, 1:1 to remove adsorbed carotenoids.

Determination of β-Carotene

The crude pigments are extracted using acetone-hexane as solvent, carotenes are separated from other pigments on magnesium oxide-supercel adsorption column, and measured at 436 nm.

Reagents:

- Acetone: alcohol free and dry, treat acetone with anhydrous Na_2SO_4, filter and distill over granular (10 mesh) of zinc.
- Hexane: Bp. 60–70 C, distilled over KOH.
- Activated magnesia and supercel (diatomaceous earth) (1+1).

Apparatus:

- Spectrophotometer
- Adsorption tube: To a 17 cm glass tube of 2 cm diameter, constricted at one end, attach a narrow (about 0.5 cm diameter) glass tube 6–8 cm long.
- Plunger: 1.25 cm long glass rod flattened at one end to fit into the adsorption tube.
- Separating funnels
- Volumetric flasks

Extraction of the Pigments:

Extract 10 g of the sample in a blender for 5 min. with 40 ml acetone, 60 ml hexane and 0.1 g magnesium carbonate. Allow the residue to settle and decant into separating funnel. Wash the residue twice with 25 ml portions of acetone, then with 25 ml hexane and combine the extracts.

Separate and remove the acetone from the extract by repeatedly washing with water. Transfer upper layer to 100 ml volumetric flask containing 9 ml acetone, and dilute to volume with hexane. If desired, alcohol may be used instead of acetone for extraction.

Chromatographic Purification

Place small glass wool or cotton plug inside the adsorption tube. Add loose adsorbent to 15 cm depth, attach tube to suction flask, and apply full vacuum of water pump. Gently press adsorbent with the plunger, and flatten the surface. The packed column should be ca. 10 cm deep. Place 1 cm layer anhydrous sodium sulphate above adsorbent. With vacuum continuously applied to flask, transfer 50 ml of the acetone-hexane extract of the sample into the column. Wash visible carotenes through adsorbent. Keep top of column covered with layer of solvent during entire operation (conveniently done by clamping inverted flask full of solvent above column with the neck 1–2 cm above surface of adsorbent).

Carotenes pass rapidly through column, bands of xantophylls, carotene oxidation products and chlorophylls should be present in column when operation is complete.

Collect entire eluate.

Determination: Transfer eluate to 100 ml vol. flask, dilute to volume with acetone-hexane (1+9) and determine carotene content photometrically at 436 nm, setting the instrument to 100% transmittance with 9% acetone in hexane. Note the carotene in the sample from the standard curve.

Standard Curve: Weigh accurately 25 mg of β-carotene p.a. (e.g. Fluka, Hoffman-La Roche or Merck). Dissolve in 2.5 ml chloroform and make up to 250 ml with 9% acetone in hexane:

1 ml = 0.1 g or 100 mg

Dilute 10 ml of this solution to 100 ml with the solvent (1 ml = 10 mg). Pipette 5, 10, 15, 20, 25 and 30 ml of the solution to separate 100 ml-volumetric flasks, each containing 3 ml acetone. Dilute to mark. The concentration will be 0.5, 1.0, 1.5, 2.0, 2.5 and 3.0 mg per ml. Measure the colour at 436 nm using 9% acetone in hexane as blank. Plot absorbance against concentration.

Calculation:

$$a = \frac{b}{d} \frac{c}{10}$$

where:

a = mg of β-Carotene/100 g
b = mg of β-Carotene per ml as read from the curve
c = Dilution
d = g of sample

Alternative Method for the Determination of Total Carotenoids and β-Carotene

Extract 10 g of the sample with 50 ml of Methanol-Petroleum Ether-Mixture (1:1) and filter through a cotton plug. Wash the residue until the washings are colourless (about 250 ml). Combine the extracts and add 25 ml water. After separating the aquatic layer wash the petroleum ether with methanol (90%) until the effluent is nearly colourless. Wash with 50 ml water to remove any residual alcohol.

Dry the petroleum ether layer (containing the total carotenoids) by filtration through anhydrous Na_2SO_4 and wash with the solvent into volumetric flask (200 ml). Determine the total carotenoids photometrically at 450 nm.

For the determination of β-carotene concentrate the petroleum ether extract under vacuum to about 2–3 ml. The carotene is separated from the other pigments on a column (25 cm long and 2 cm in diameter) containing aluminium oxide partially inactivated with 8% water. Place small glass wool or cotton plug inside the adsorption tube and add petroleum ether and then the aluminium oxide. Transfer the concentrated extract to the column and wash the visible carotenes through adsorbent. Keep top of column covered with layer of solvent during entire operation. After the first 25 ml eluate, the carotene fraction will be at the end of column. Collect the next 25 ml in a volumetric flask and determine the carotene content at 450 nm according to the standard curve [9].

2.12 Determination of Anthocyanins

The anthocyanin undergoes a structural transformation with changes in pH. Anthocyanin pigments can be described as being indicators, i.e. their hue (shade or colour) and intensity (depth of colour) change with pH. At pH 1.0, anthocyanins exist in the high coloured oxonium or flavilium form and at pH 4.5 they are predominantly in the colourless carbinol form. The quantitative procedure for determining anthocyanin which will be described is based on these facts [31].

One aliquot of an aqueous anthocyanin solution is adjusted to pH 1.0 and another to pH 4.5. The difference in absorbance at the wavelength of maximum absorption (between 510 and 540 nm, for many products it will be 520) will be proportional to anthocyanin content.

Buffers:

- pH 4.5 buffer: 400 ml of 1 M Sodium acetate (136 g/l) + 240 ml of 1 N HCl (83.0 ml conc. HCl/l) + 360 ml distilled water.
- pH 1.0 buffer: 125 ml of 0.2 N KCl (14.9 g/l) + 385 ml of 0.2 N HCl.

The pH of the buffers can be adjusted as required to obtain final pH values of 1.0 and 4.5.

Determination: To 20 g of the sample add 80 ml of buffer and mix in a blender at full speed. The order of dilution must be such that the sample at pH 1.0 will have an absorbance of less than 1.0 and preferably in the range of 0.4–0.6. The dilution strength should be the same for both 1 and 4.5 samples.

The diluted samples should be clear and not contain any haze or sediment. Any sediment should be removed by centrifuging and filtrating the sample. If the sample is free of haze, the absorbance at 700 nm should be 0.

Turbidity (haze) can be corrected for by measuring the absorbance at 700 nm and subtracting this from the absorbance at the wavelength of maximum absorption (e.g. 510 nm):

$$A = (A_1 - A_2) - (A_3 - A_4)$$

$$B = \frac{A}{EL} \ 10^3 \ MW \ d$$

where:

A = Adsorbance difference between pH 1.0 and 4.5
A_1 = A_{510nm} at pH 1.0
A_2 = A_{700nm} at pH 1.0
A_3 = A_{510nm} at pH 4.5
A_4 = A_{700nm} at pH 4.5
B = Concentration mg/l
EL = Molar absorbance of the major anthocyanin (Table 5).
MW = Molecular weight of the major anthocyanin in the fruit (Table 5).
d = Dilution factor

It should be emphasized that the pH differential method is a measure of the monomeric anthocyanin pigments and the result may not seem to be correlated with the colour intensity of the fruit product as it is judged visually. This is because polymeric anthocyanins and brown pigments arising from enzymatic browning, the Maillard reaction and anthocyanin degradation also contribute to the colour intensity.

Colour Density: The colour density of a fruit product can be determined by adding the absorbance of the control sample at 420 nm (Browning) and at the anthocyanin absorbance maximum (for many products this will be 520 nm). Turbidity can be corrected for by substracting any absorbance at 700 nm.

$$CD = (A_1 - A_2) + (A_3 - A_2)$$

where:

CD = Colour Density
A_1 = Absorbance at 520 nm
A_2 = Absorbance at 700 nm
A_3 = Absorbance at 420 nm
d   = Dilution factor

Colour Degradation Index:
The colour degradation index can be determined by dividing the absorbance at the anthocyanin absorbance maximum (at 520 nm) with the absorbance at 430 nm (degradation products).

$$DI = [(A_1 - A_2) / (A_3 - A_2)] \, d$$

where:

DI = Colour Degradation Index
A_1 = Absorbance at 520 nm
A_2 = Absorbance at 700 nm
A_3 = Absorbance at 430 nm
d   = Dilution factor

2.13 Determination of Benzoic Acid

Sodium benzoate (and nowadays also potassium benzoate) is the usual preservative for fruit juices, pulps and jams. Imparting its maximum activity in the pH range of 2.5–4.0, benzoate is more effective in controlling yeast and bacteria and less effective in controlling moulds. It is suitable for use in fruit products with a pH below 4.0 or in products that may be acidified to this range. Potassium benzoate evolved in response to consumers' interest in reduced sodium intake. Like the sodium salt, it has a sweetish and an astringent taste.

In a NaCl solution of the sample, the benzoic acid present in the sample is converted into watersoluble sodium benzoate by the addition of NaOH. When the solution is acidified with excess HCl, water *in*soluble benzoic acid is formed, which is extracted with chloroform. The chloroform is removed by evaporation and the residue containing benzoic acid is dissolved in alcohol and then titrated with NaOH.

Reagents:

- 0.05 N NaOH (2.075 g/l)
- NaOH, 10% solution
- HCl (Dil. 1+3 acid + water)
- NaCl, powder and saturated salt solution
- Ethanol (neutral to phenolphthalein)
- Phenolphthalein indicator (1 g in 100 ml ethanol)

Determination: To 100 g of sample add 15 g NaCl and transfer the mixture to a 500 ml volumetric flask rinsing with 150 ml of saturated salt solution, make alkaline testing with litmus paper with 10% NaOH solution, and dilute to mark with saturated salt solution. Shake thoroughly, allow to stand for at least 2 h with frequent shaking, filter through Whatman No. 4 paper.

Pipette 100 ml of the filtrate into a 500 ml separating funnel. Neutralize using litmus paper as indicator with dil. HCl (1 + 3) and add 5 ml HCl in excess. Extract with chloroform using successive portions of 70, 50, 40 and 30 ml. Draw off as much clear chloroform solution as possible after each extraction. The chloroform layer can be also washed with water.

Transfer the combined chloroform extracts to a 250 ml flask. Distil very slowly at low temperature to about one-fourth of the original volume, and evaporate to dryness at room temperature in a current of dry air on a water bath till only a few drops remain. Dry the residue overnight in a desiccator containing H_2SO_4. Dissolve the residue in 50 ml of alcohol (neutral to phenolphthalein), and add 12–15 ml of water and 1–2 drops of phenolphthalein indicator and titrate against 0.05 N NaOH.

Calculation:

$$1 \text{ ml of } 0.05 \text{ N NaOH} = 0.0061 \text{ g of benzoic acid}$$
$$= 0.0072 \text{ g of sodium benzoate}$$

$$a = \frac{b\,c\,122\,d\,1000}{e\,f}$$

where:

- a = ppm Benzoic acid
- b = Titration
- c = N of NaOH
- d = Volume made up
- e = Volume taken for estimation
- f = Wt of the sample

HPLC-Determination: This method is recommended for numerous routine analyses of benzoic acid and if a HPLC-Apparatus is available. The Beckman model 330 with model 153 detector is recommended, or an isocratic HPLC with fixed wavelength (254 nm) detector and recorder. A column (25 cm x 4.6 mm) with Partisil 10 μ ODS 2 is required. The HPLC-Instrument should be used according to the manufacturer's instructions. A pre column of the same material is recommended [10, 32, 33, 34].

Reagents: HPLC Mobile Phase: Methanol: Water: Perchloric acid (35:64:1). Mix 350 ml of methanol (A.R. or HPLC grade) with 600 ml distilled water and add 10 ml of perchloric acid (A.R. grade 60% w/v) and mix gently. Dilute to 1000 ml with water and mix again. Degas the mobile phase by ultrasonic or helium purge before use (or by reflex at 80 C for 1–2 min).

Alternative mobile phase: Methanol: Water: Acetic acid (45:53:2). Mix 450 ml of methanol with 500 ml of water. Add 20 ml of acetic acid and mix gently and dilute to 1000 ml with water and mix again.

Benzoic Acid Standard: Stock solution (1000 ppm): dissolve 1.00 g of dry benzoic acid (A.R. grade) in 10–20 ml dilute sodium hydroxide solution. Dilute to 1000 ml with water.

Working solution (100 ppm, it must be prepared daily): pipette 10.0 ml of the stock solution into a 100 ml volumetric flask and add 20 ml distilled water and mix. Add 50 ml of methanol and mix. Dilute to 100 ml with water.

Determination: Weigh 5 g of the sample in a 50 ml volumetric flask and add 20 ml of water. Shake to disperse the sample. Add 25 ml of methanol, shake and dilute to 50 ml with water. Centrifuge the content at 3000 rpm for 5 minutes. The supernatant liquor is suitable for injection.

Inject 20 μl of the benzoic acid standard solution and record the peak height of duplicate injection. Inject 20 μl of the prepared sample and record the peak height for benzoic acid.

$$a = \frac{b\,100\,c}{d}$$

where :

a = ppm Benzoic acid
b = Peak height in sample
c = Dilution factor
d = Peak height in standard solution

More information about HPLC determination of benzoic acid is given by Macrae [10]. Woodward et al. [33] and Hewlett-Packard [34].

2.14 Determination of Sorbic Acid

Sorbic acid is the only unsaturated organic acid permitted as food preservative. Sorbic acid and potassium sorbate are more effective against yeast and moulds than against bacteria. Unlike benzoic acid which is effective to maximum pH value of 4.5, sorbic acid is active to a maximum of pH 6.5. With neutral taste characteristics, potassium sorbate is an important preservative in fruit juices, essences, soft drinks and jams, in contrast to benzoates which may modify the flavour profile of these products.

The determination is based on the steam distillation of sorbic acid from sulphuric acid after saturating the sample with magnesium sulphate. The sorbic acid in the distillate can be determined either by UV-Absorption at pH 2–4, or by oxidation with acid dichromate to malonaldehyde, which reacts with thiobarbituric acid to give a red colour having a maximum absorption at 532 nm.

Reagents:

- *Sulphuric acid 2 N*, dilute 14.2 ml H_2SO_4 with H_2O to 250 ml.
- *Sulphuric acid 0.3 N*, dilute 15 ml 2 N H_2SO_4 to 100 ml.
- *Potassium dichromate solution*, dissolve 147 mg $K_2Cr_2O_7$ in water and dilute to 100 ml.
- *Thiobarbituric acid solution (0.5% w/v)*, dissolve 250 mg thiobarbituric acid in 5 ml 0.5 N NaOH in a 50-ml flask by swirling under hot water (0.5 N NaOH = 20 g/l). Add 20 ml water, neutralize with 3 ml 1 N HCl, and dilute to 50 ml with water. The solution must be freshly prepared daily.
- *Sorbic acid standard solution (0.1 mg/ml)*, accurately weigh 134 mg potassium sorbate (equiv. to 100 mg sorbic acid) and dilute to 1 l with water. The solution is stable for several days when refrigerated.

Determination: Add 2 g of sample to 10 ml 2 N H_2SO_4 and 10 g $MgSO_4$. 7 H_2O, and steam distil rapidly. Do not attempt to apply any heat directly to the flask containing the sample during distillation, otherwise a coloured distillate may result. Collect 100–150 ml distillate in a 250-ml flask within approx. 45 min. Rinse condenser with water and dilute distillate to 250 ml with water and mix. The sorbic acid can be determined by UV-absorption at 363 and 280 nm using water as a reference. Make a suitable dilution of distillate with 0.3 N sulphuric acid.

Colorimetric determination, pipette 2 ml of distillate into 15 ml test-tube and add 2 ml of freshly prepared acidified dichromate solution (1 ml 0.3 N H_2SO_4 and 1 ml $K_2Cr_2O_7$ solution) and heat in boiling water exactly 5 min. Immerse tubes in beaker of cold water and add 2 ml thiobarbituric acid solution. Replace in boiling water bath and heat additional 10 min. Cool and determine absorption at 532 nm against blank, using matched 1 cm cells. Determine concentration of sorbic acid from standard curve and calculate concentration in sample.

Preparation of the standard curve, just before use, pipette 5, 10 and 15 ml sorbic acid standard solution into 500 ml vol. flasks. Dilute each to vol. and mix. Pipette 2 ml of each solution and 2 ml (for blank) into 15 ml test-tubes. Add the dichromate solution and proceed as in the above-mentioned method.

2.15 Determination of Sulphur Dioxide

Sulphur dioxide is the only permitted inorganic preservative. It is generally used in the form of sulphites (sodium metabisulphite $Na_2S_2O_5$), but for the purpose of the regulations the amount present is calculated as sulphur dioxide, SO_2. SO_2 is useful in controlling microbial growth and maintaining oxidative stability in dehydrated fruits, fruit juices, syrups, concentrates and purees. It assists in conserving vitamin C, but appears to inactivate vitamin B_1. Because the safety of SO_2 and the wisdom of using it have been the subject of recent questions and public concern, its use has been limited in many countries.

The determination method is based on the distillation of SO_2 in the presence of acid (and in an inert atmosphere), absorption of SO_2 in an oxidising agent (iodine solution), which converts the sulphurous acid to sulphuric acid, and estimation by titration with standard sodium thiosulphate solution. The method determine the total (free + combined) SO_2 and is sufficiently rapid. If it is desired to determine the free SO_2 only, use the same procedure but do not add any acid. The distillation apparatus used (Fig 7) is the same as that employed in Kjeldahl ammonia distillation.

Reagents:

- HCl (diluted 1:4 acid : water)
- *0.1 N iodine solution*, dissolve 13.5 g of pure resublimed iodine in a solution of 24 g of potassium iodide in 200 ml of water, and dilute to 1 liter. Standardize the solution by titrating against a known volume of standard thiosulphate solution using starch as indicator.
- *Starch indicator* (0.5%), mix 0.5 g of soluble starch in 15 ml cold water and pour into 100 ml of hot water, boil for 2 min.
- *0.1 N standard sodium thiosulphate solution* (24.82 g of $Na_2S_2O_3 . 5 H_2O/l$), dissolve 25 g of sodium thiosulphate in 200 ml water, and dilute to 1 liter. Mix thoroughly, allow to stand for a few days, and then siphon off the clear liquid. Standardize the solution with potassium dichromate ($K_2Cr_2O_7$). Weigh 0.20 to 0.23 g of K-dichromate (dried for 2 h at 105 C). Transfer to a 250 ml beaker using about 150 ml of water. Add 2 g of potassium iodide and mix. Add 20 ml of 1 N HCl, swirl and allow to stand for 10 min. Start titrating with the sodium thiosulphate from burette, adding about 80% of the required

amount. Add 1 ml of starch and complete titration to a point, where the solution changes from blue-green to light green.

$$a = \frac{b}{c\,0.049037}$$

where:

a = N of standard sodium thiosulphate solution
b = Wt of potassium dicromate (g)
c = Volume of sodium thiosulphate solution (ml)

Determination: Boil about 500 ml of water in the flask as shown in Fig. 7. Take 50 g of the sample in the Kjeldahl flask and connect to the apparatus. Add 200 ml of HCl, turn on the burner and boil as quickly as possible. Allow the distillate to pass down the condenser and collect in the vessel containing 0.1 N iodine solution. Most of the sulphur dioxide is usually evolved within 5 min boiling. Continue the distillation for another 5 min., and titrate the excess of iodine with 0.1 N sodium thiosulphate. A blank determination with no sample should be done.

Calculation:

1 ml of 0.1 N iodine = 1 ml of 0.1 N thiosulphate = 3.2 mg of SO_2

$$a = \frac{(b-c)\,320}{d}$$

where:

a = SO_2 mg/100 g
b = Sample titrate
c = Blank
d = Wt of sample

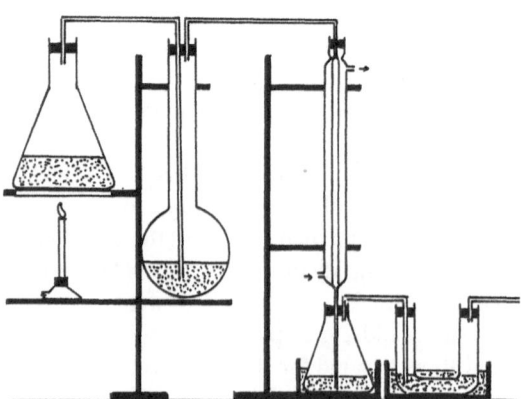

Fig. 7. Apparatus for SO_2 distillation

2.16 Determination of Pectinesterase Activity

Determination of Pectinesterase Activity

Pecitc substances are composed mostly of polygalacturonic acid of varying degrees of esterification and neutralization. Because tropical fruits contain considerable amounts of pectin substances and contain acid and sugar, one can readily understand why fruit pulp under some conditions will tend to have gelation or clouding instability.

There are several methods available for analyzing pulps and concentrates to find what possible degree of gelation or of clouding instability that they may possess. One such method determines the activity of pectinesterase, which degrades pectin by removing methoxyl groups from the pectin molecule thus leaving free caboxyl groups. This makes pectin susceptible to precipitation by calcium or other polyvalent cations (clouding instability or calcification) and contributing to gelation in the fruit products. As the pectinesterase activity increases the tendency towards gelation of juices and concentrates increases.

In general, the optimum pH range for pectinesterase (EC 3.1.1.11) from plants is 7 to 8 compared with 4 to 5 for mould pectinesterase. Pectinesterase, especially from tropical fruit juices, is relatively heat-stable. Therefore, the determination of its activity provides valuable information about heat-treatment. This test may be applied to concentrates, evaporator feed juice, cut-back juice, and may be used to measure the efficiency of the pasteurizer.

Pectinesterase activity can be determined titrimetrically by estimating the free carboxyl groups formed in pectin as a result of enzyme action. The amount of 0.05 N NaOH required to maintain the pH of the substrate solution at 7.0 at 30 C, can be measured using an automatic titrator.

Additional pectin is added to the reaction mixture to supply an excess of substrate for the enzyme to hydrolyze. The pectinesterase is usually adsorbed on the insoluble cellular solids. It is, therefore, necessary to extract the plant material in 0.25 M salt solution and maintain a pH 8 for about 1 h while the enzyme dissolves [35].

Reagents:

- Commercial apple pectin (Obipectin, green ribbon, DE 62.0%)
- Sodium chloride stock solution, 2 M: dissolve one molecular weight (58.45 g) NaCl in 500 ml water.
- Sodium hydroxide 0.05 N (2.075 g/l) 0.10 N (4.15 g/l) 0.3125 N (12.97 g/l)
- Anti-foam, diglycol stearate.

Equipment:

- pH-Meter
- Waring blender
- Stop watch

- Automatic titrator or Burette, 10 ml and Micro-burette, 10 ml
- 1-l volumetric flask
- Magnetic stirrer with teflon covered stirring rods.

Preparation of substrate solution (1% pectin-salt solution): Place 75 ml 2 M NaCl solution in the 1-l volumetric flask, and fill to mark with water. Fill blender jar approx. 3/4 full with this solution, run blender and add slowly 10 g of pectin, add remainder of salt solution, blend approx. 1 min, and place in refrigerator.

Determination:

- Place 20 ml of the juice, 12 Brix, in a 250-ml beaker, titrate to phenolphthalein and point with 0.3125 N NaOH.
- Add 40 ml of the 1% pectin solution.
- Adjust pH with 0.1 N NaOH to 7.0, constantly stir with mixer, add antifoam (small pinch) and allow pH to return 7.0
- Add with micro-burette 1 ml of 0.05 N NaOH, and start time at closing of burette. Measure time for pH to readjust to 7.0 (and volume of alkali required to maintain the pH at the constant value).
- Repeat adding second portion of alkali and get average time.

Calculation: Pectinesterase-Units (PEU) represent the milliequivalent of ester hydrolyzed per minute per ml of the sample, or one unit of PE is the amout of enzyme which liberates 1 μmole of carboxyl groups per minute.

$$a = \frac{b\,c\,10^4}{d\,e}$$

where:

a = Pectinesterase-Units PEU.104
b = ml of NaOH
c = N of NaOH
d = minutes
e = ml of the sample

Example: if the N of the NaOH is exactly 0.05 and the average time for the action was 15 min, then the

$$PEU \times 10^4 = \frac{1\text{ ml} \times 0.05 \times 10^4}{15 \times 20} = 1.67$$

2.17 Measurement of Cloud Stability

The serum pectin present in a water soluble form in fruit juices. More important is the insoluble pectin in the cloud particles, because its effect on the cloud stability of the juice. Cloud particles (haze) remain in suspension normally due to their negative charge and to the surrounding aqueous capsule. Both of these factors are dependent on the pectin proportion which at pH less than 4.0 surrounds the positively charged protein/carbohydrate proportion of the haze-forming particle.

Modification or elimination of the pectins negatively effects the cloud stability. When the level of esterification of the serum pectin is reduced by the action of pectinesterases, the water solubility falls and reactivity to calcium increases. At a pH lower than 3.2, serum pectin precipitates as calcium pectinate together with the cloud particle.

Polygalacturonases (hydrolases) cut down and reduce the negatively charged pectin envelope which surrounds the cloud particle. This will, at a pH lower than 4.0, in some areas of the particle free the positively charged protein/carbohydrate parts of the cloud particle and will lead to heterogeneous charged areas on the surface. Coagulation and sedimentation are the result of these charge separations. Where residual activities of pectinesterases and polygalacturonases are present in juice, pulps and drinks, a precipitation of cloud material may take place between pH 2.5 and pH 4.0. Adequate heat treatment of the raw materials is necessary to prevent cloud loss during storage or subsequent clarification. Cloud retention and heat treatment was systematically studied in citrus juices and recently also in tropical fruit juices. Pectinesterase is known to be the most heat resistant enzyme present in fruit juices. Even after heat processing, some residual pectinesterase activity may remain in pasteurized products. Heat resistance of Lactobacilli and yeasts has been found to be lower than that of the heat resistant pectinesterase. We found the inactivation time of pectinesterase in mango and guava pulp at 97 C is about 1 min, and we recommend the determination of thermal process schedule for tropical fruits on the basis of inactivation of pectinesterase.

Because most of the colour pigments and aroma components are localized in the cloud particles, cloud loss or separation leaving a clear supernatant layer is a serious quality defect which reduces the attractiveness and marketability of the product.

Two methods are available for measuring the cloud stability of fruit juices and nectars [36]. The first is the centrifugation method, which is commonly used to measure the cloud stability of citrus juices. This involves centrifugation of a 10 ml sample in a conical centrifuge tube for 10 min at 360 x g and monitoring the turbidity of the supernatant by measuring the extinction (E) at 660 nm in a spectrophotometer.

The second method is the recommended method. The nectar is placed in a 100-ml serum bottle and allowed to stand undisturbed at 20 C. The height of the cloudy phase during sedimentation is recorded daily in the first week and every two or three days thereafter. The degree of stability is to be measured as a percentage of the height in respect to the total height of the nectar (100 ml), expressed as follows:

$$a = \frac{b\ 100}{c}$$

where:

a = Degree of stability
b = Height of cloudy phase
c = Total height

Stevens Cloud Stability Test [36]

This test was developed for pasteurized citrus concentrates and is used as an indicator of cloud stability of other fruit juices. The Stevens test is a qualitative determination for residual enzyme activity.

Equipment:

– pH-Meter
– Balance
– Water-bath at 48 C
– Screw cap test tubes, 250 ml beaker, 100 ml and 1000 ml graduated cylinder.

Reagents:

– *Pectin solution 2%:* 20 g rapid-set pectin and 1.5 g sodium benzoate up to one litre.
– *Citric acid solution 50%:* 500 g up to one litre water.
– *Barium chloride solution 20%:* 200 g $BaCl_2$. 2 H_2O up to one litre water.
– *Sodium benzoate solution 40%:* 400 g up to one litre water.

Determination:

– Measure 75 ml water into 100 ml cylinder and add 15 ml fruit pulp. Mix well and pour into 250 ml beaker.
– Add 5 ml of 2% pectin solution and mix well.
– Add 2 ml of 20% barium chloride solution and mix well.
– Add 1 ml of 40% sodium benzoate solution and mix well.
– By means of the pH-Meter adjust acidity of the mix with 50% citric acid solution to a pH between 3.0 and 3.1.
– Fill screw cap test tubes and incubate at a temperature of 48 C.

– Examine for indication of clear supernatant liquor at 12, 24, 48, and 72 hours intervals.

Remark: The pectin that should be used will not form a gel when 0.1 g pectin is dissolved in 100 ml water and 1.3 ml of barium chloride solution is added.

2.18 Colour Index

Application: Mango and Passion Fruit (Juice, Purée, Concentrate and Powder).

Principle of the Method: A total petroleum ether-soluble and water-soluble colouring matter determination. The petroleum-ether-soluble colouring matter is a measure of the true carotenoid content of the sample. The water-soluble colouring matter gives an indication of the amount of browning reaction products formed during processing (heat damage) and storage.

The petroleum ether and water-soluble fractions are extracted from the sample by shaking the sample in the presence of N,N-dimethyl formamide (to give complete carotenoid extraction and cellular breakdown) and sodium chloride (to prevent the formation of emulsions). The colouring matters are measured spectrophotometrically and expressed as the extinction coefficient $E^{1\%}/_{1cm}$.

Reagents:

– Petroleum ether, bp 80–100 C
– N,N-dimethyl formamide Sodium chloride.
– 2% m/v aqueous solution, filtered through Whatman No. 42 paper

Apparatus: Extraction apparatus consisting of a 250-ml, wide-mouthed, flat-bottomed flask, a laboratory stirrer and an anti-splash cone. Spectrophotometer Cuvettes, U/V grade silica, optical path length 10 mm Filter papers, Whatman 42 and 542

Determination Procedure

Blank Test: Measure 50 ml petroleum ether, 3 ml N,N-dimethyl-formamide and 47 ml sodium chloride solution into the extraction flask. Stir with the blades about 5 mm from the flask bottom for 5 minutes. Transfer the total mixture to a 100 ml cylinder and allow to separate into two phases. Remove by pipette 20 ml of the upper petroleum ether layer and filter through a plug of cotton wool, discarding the first 5 ml. Read absorbance of the filtrate in a 1 cm cell at 450 nm against petroleum ether as blank. The filtrate should show no absorbance.

Remove by pipette 20 ml of the lower aqueous phase and filter through a combined Whatman 542 and 42 filter paper, discarding the first 5 ml. Read the absorbance of the filtrate in a 1 cm cell at 380 nm against water as a blank. The absorbance should be less than 0.002.

Sample test: Weigh accurately a quantity of sample equivalent to 0.1 g mango solids into the extraction flask. Add 3 ml *N,N*-dimethyl formamide and shake gently for 10 minutes. Add 50 ml petroleum ether and 47 ml sodium chloride solution. Stir the mixture gently for 2 minutes then rapidly for 5 minutes. Transfer the mixture to a 100 ml cylinder and allow to settle, if necessary, centrifuge to separate the two layers. Remove by pipette 20 ml of the petroleum ether layer and filter through a Whatman 42 paper, discarding the first 5 ml. Read the absorbance of the filtrate in a 1 cm cell at 450 nm against petroleum ether as blank. Remove by pipette 20 ml of the aqueous phase through a combined Whatman 542 and 42 filter paper and discard the first 5 ml. Read the absorbance of the filtrate in a 1 cm cell at 380 nm against water as a blank.

Calculation: Calculate the extinction coefficient of both layers as follows:

Let:
– Absorbance   = A
– Mango solids in sample (%) = T
– Mass of the sample (g)   = M

$$\text{Then: Extinction E } 1\%/1\text{cm} = \frac{50\,A}{T \times M}$$

2.19 Determination of Non-enzymatic Browning

Heat processing and subsequent storage of fruit products is often accompanied by an unattractive discoloration due to reaction of sugars with amino substances, through non-enzymatic browning. Three factors, namely, processing condition, storage time and temperature influence the colour quality of the product. Non-enzymatic browning is considered one of the major causes of quality loss of fruit products (see chapter Deterioration and Spoilage of fruit products).

The colouring compounds have not been identified, but they are known to be water-soluble. The increase in absorbance of a sample extract at 440 nm (or at 420 and 380 nm) is taken as a measure of non-enzymatic browning [38–40].

Determination Procedure: Juices, remove juice pulp and particles by centrifugation at 2000 rpm. Dilute the liquid with an equal volume of 95% ethanol. Centrifuge again and filter through Whatman No. 1 paper.

Dried fruits, extract 10 g of sample with 100 ml 60% ethanol for 12 h and filter. For samples containing chlorophyll, shake the alcoholic extract with three lots of 50 ml benzene. If the filtrate is not clear filter using filter aid or through millipore filter.

Measure the colour at 440 nm or at 420 nm (of if it is possible, measure the spectrum between 380 and 440 nm) with a spectrophotometer using 60% aqueous alcohol as blank.

2.20 Determination of Furfural

Furfural is formed during the processing of fruits (heating, evaporation, drying) and during storage at relatively high temperatures. It is thought to be a breakdown product of ascorbic acid and can be also derived from the decomposition of pentoses.

A strong relationship between flavour degradation of fruit products and the formation of furfural has been observed. But furfural itself does not cause the off-flavour of temperature abuse of fruit products.

The increase in furfural content is regular and depends on temperature and time, runs parallel with the flavour alteration and its determination is relatively simple (in comparison with gas chromatographic analysis of flavour compounds). Therefore, furfural has been recommended as an useful indicator of temperature abuse in commercially produced fruit products.

Furfural can be determined after distillation by a colorimetric method based on its reaction with aniline and acetic acid. The addition of $SnCl_2$ in HCl improves colour intensity and stability. The measurement of absorbance at 515 nm is specific for furfural. Hydroxymethyl-furfural and methylfurfural absorb at different wavelengths and do not interfere with the results [41].

Reagents

- *Purified aniline*, one or two distillation of aniline may be required to obtain a colourless product. Under refrigeration it remains colourless several months.
- *Stannous chloride stock solution (20%)*, dissolve 2 g $SnCl_2$. 2 H_2O in 6 N HCl (1 + 1). Best results can be obtained for colorimetric analysis of furfural using 1% $SnCl_2$ and 0.6 N HCl, with which maximum colour was reached at 50 min and persisted for 25 min.

– *Stannous chloride-aniline-acetic acid,* add 1 ml 20% $SnCl_2$ stock solution during dilution of 2 ml aniline with glacial acetic acid to 20 ml total volume.
– *Furfural standards,* weigh 100 mg furfural into 100 ml water (1 mg/ml). Dilute the solution to 100-fold (10 µg/ml). Dilute the second solution 5-, 10-, and 20-fold to obtain furfural concentration of 2, 1, and 0.5 µg/ml.

Determination [41]: Place 200 ml fruit juice (or 50 g of ground dry fruit in 150 ml water) into a 500 ml boiling flask and connect to steam distillation. Add 1 ml silicone antifoam and a few boiling chips. A Scott-Veldhuis oil distillation apparatus can also be used.

Collect 50 ml of distillate in an ice-chilled graduated cylinder. Do not rinse condenser between replicates, but do rinse with water and methanol and blow dry with air between different samples. After thorough mixing, pipet a 2-ml aliquot into a test tube. Add 2 ml of 95% ethanol and 1 ml of the aniline-acetic acid reagent. Distillates containing more than 2 µg/ml furfural require dilution with water. After the addition of the reagent and mixing add glacial acetic acid to make the total volume 20 ml. Allow the mixture to stand for 10 min and read the absorbance of the mixture at 515 nm with a spectrophotometer, using circular tubes with effective 1 cm light path.

Prepare a calibration curve using a series of working standards containing 2, 1, and 0.5 µg/ml furfural.

Furfural recovery studies must be made to determine the distillation efficiency. Prepare 600 ml each of standards containing 25, 50, 75, and 100 µg per 200 ml from the secondary standard. The dilution can be made with water or juice. The 600 ml provides for three distillations of 200 ml each.

Calculation:

$$a = \frac{b}{c}$$

where:

a = Distillation efficiency
b = µg furfural/200ml recovered
c = µg furfural/200ml

$$\text{ppb furfural (µg/l)} = \frac{125 \times F \times A}{DE}$$

where:

F   = absorbance factor as µg furfural per absorbance unit
       (determined from the slope of the calibration plot, usually 0.122)
A   = absorbance sample
125 = dilution factor when 2 ml distillate is used
DE = Distillation efficiency, usually 0.34

HPLC-Determination of Furfural [42]

The current method of furfural analysis uses a colorimetric reaction between furfural and aniline in the presence of glacial acetic acid and stannous chloride. This method is time-consuming, taking approximately 1 h for colour development and requires the hazardous chemical aniline. Marcy and Rouseff [42] developed a HPLC method that is more sensitive, is less time-consuming, and utilizes less hazardous materials. The minimum detectable level of furfural is 2 ppb, and a linear response is obtained from 12 to 20 000 ppb. The method is simple and well suited to routine analysis.

Sample preparation: Furfural is separated from most of the other juice components using the distillation procedure previously described. The first 10 ml of distillate was collected (5% of juice volume), mixed, and used for analysis. The average recovery of furfural ($N = 30$) from spiked orange juice samples was 37.7 $\pm$ 0.7%.

Chromatographic Conditions: Methanol-water (70:30 v/v) and a Du Pont Torbax ODS column (Wilmington, DE, USA), 4.6 mm i.d. x 25 cm at 1 ml/min separated furfural in 8.5 min.

Chromatographic hardware consisted of a Waters Associates (Milford, MA, USA) m-6000A pump, U-6K injector, Model 440, fixed wavelength detector with the 280 nm aperture kit, and a Spectra-Physics SP4270 recording integrator.

Concentration factors were obtained from furfural standards (reagent-grade furfural was doubly distilled before use) and used to calculate concentration (the relationship between peak area and furfural concentration). Calibration standards of 0.5, 1.0, and 2.0 ppm of furfural in the distilled sample are used in the colorimetric procedure.

Juice samples that have more than 300 ppb require dilution prior to analysis.

2.21 Determination of Hydroxymethylfurfural

Hydroxymethylfurfural (HMF) is, like furfural, an intermediate product of the Maillard-reaction. HMF can be formed in fruit juices directly by acid-catalyzed sugar, mainly hexoses, degradation. The formation of HMF begins even before any visible browning or alteration in taste occur. HMF may polymerize or combine with amino compounds to give rise to non-enzymatic brown colour.

In untreated juices, the HMF content is very low (practically 0). A HMF-content of more than 5 mg/l in juices indicates a loss of quality and more than 10 mg/l an unsuitable working technology or long storage at high temperatures.

The determination of HMF in juices, marmalade, honey and fruit syrup is based on the reaction of HMF with barbituric acid and *p*-toluidine to give a red-coloured substance which can be measured at 550 nm. The photometric determination of HMF in fruit juices is recommended by the INTERNATIO-NAL FEDERATION OF FRUIT JUICE PRODUCERS [9] and in Honey by the CODEX ALIMENTARIUS COMMISSION of the FAO/WHO. Small amounts of HMF occur naturally in honey due to acid-catalysed dehydration of the hexose sugars and there is an increase in HMF content during the storage. The international standard for honey (FAO/WHO 1987) permit no more than 80 mg HMF/kg honey. Higher values indicate an adulteration with artificial products.

Apparatus:

– Analytical balance
– Spectrophotometer to read at 550 nm
– Volumetric flasks, 50, 100 and 1000 ml
– Test tubes

Reagents:

– *Barbituric acid solution:* transfer 500 mg barbituric acid to a 100 ml volumetric flask using 70 ml water. Place in a hot water bath until dissolved, cool and make up to volume.
– *p-Toluidine solution:* dissolve 10 g toluidine in 50 ml isopropanol by gentle warming on a water bath. Transfer to a 100-ml volumetric flask with isopropanol and add 10 ml glacial acetic acid. Cool and make up to volume with isopropanol.
  Keep the solution in the dark and do not use for at least 24 hours. Beware: carcinogenic reagent.
– *Oxygen free distilled water:* pass nitrogene gas through boiling water and cool.
– *Carrez-solution I:* dissolve 150 g potassium ferro-cyanide ($K_4Fe(CN)_6 . 3 H_2O$) in up to 1 litre water.
– *Carrez-solution II:* dissolve 300 g zinc acetate ($Zn(Ch_3COO)_2 . H_2O$) in up to 1 litre water.
– *Active carbon* (Claro carbon F).
– *Iodine solution:* 0.1 N
– *Starch solution:* 0.5% (w/v)

Procedure: Treat 25 g juice or diluted concentrate or marmalade to adjust the pH to 7–8. Add some drops of isopropanol to reduce foam formation. Add 4 ml of 0.5% starch solution and add by drops 0.1 N iodine solution until the blue colour of iodine-starch remains for 15–20 s.

Transfer the solution to 50 ml volumetric flask and add 1 ml Carrez-solution I and of Carrez-solution II. Make up to volume with water and filter the solution.

Take 2 ml of filtrate into each of two test tubes, add 5 ml of toluidine solution to each. Into one test tube pipette 1 ml water and into the other 1 ml barbituric acid solution and shake both mixtures. The one with added water serves as the water blank. The addition of the reagents should be done without pause and should be finished in about 1–2 min.

Measure the absorbance of the sample against the blank in a 1 cm cell at 550 nm as soon as the maximum value is reached.

Calculation: The method may be calibrated by using a standard solution of HMF after spectrophotometric assay at 284 nm (molar extinction coefficient = 16,830) using 0.300 µg of the standard. An approximate figure can be obtained from the equation:

$$a = \frac{b\ 19.2}{c}$$

where:

a = HMF mg/100 g
b = Absorbance
c = Cell path length

HPLC Determination of Hydroxymethylfurfural [43–46]

The photometric determination of HMF has been criticised for the instability of the developed colour, the temperature dependence and the use of a carcinogenic reagent.

Recently, a HPLC-method has been developed overcoming these problems. The method is based on a sample preparation method for the simultaneous determination of furfural and HMF using a disposable C-18 cartridge [43].

Preparation of test solutions and cartridges: A stock 0.1% (w/v) standard solution of furfural and HMF was prepared on the day of use with 10% methanol. Furfural was doubly distilled before use, but HMF was used without further purification. The Carrez clarification reagent consisted of 15% (w/v) of Carrez I (potassium ferro cyamide) and 30 % (w/v) Carrez II (zinc sulphate) solution. Both were prepared with distilled water. A short disposable Sep-PAK C-18 cartridge (Waters Associates, Milford, MA, USA) was pre-wetted with 2 ml of methanol followed by 5 ml of water before use.

Isolation of furfural and HMF: Ten gram of juice was pipetted into a 50 ml centrifuge tube and 0.5 ml of each Carrez I and II solution was added slowly with gentle mixing. After standing for 5 min, the mixture was centrifuged at 2000 x g for 5 min. One milliliter of clear supernatant juice was pipetted into a syringe and passed through the conditioned C-18 cartridge. After washing the cartridge with

0.5 ml of hexane, the furfurals were eluted twice with 3 ml of ethyl acetate, and dried with anhydrous sodium sulphate. Eluates were filtered through a 0.45 μm disc filter before injection.

Determination: A Waters Model 6000A pump, Model U6K injector and Spectromonitor D were used. A RCM 100 radial compression module with a Radial-RAK C-18 column (100 x 8 mm i.d.), and Resolve C-18 pre-column from Waters were used for the analytical system. Analysis was carried out by injecting 20 μl of sample or standard onto the column. Acetonitrile-water (15:85, v/v) was used isocratically at 2 ml/min.. The effluent was monitored at 280 nm. Retention times and peak areas were determined with a Spectra-Physics Model SP4270 computing integrator. The lower limit of detection for both compounds was 50 ppb. The recovery studies by comparing chromatographic peaks areas from fresh orange juice with known amount of furfural added and HMF (50 to 100 000 ppb) against peak areas generated by direct injection of standard solutions of furfural and HMF, ranged between 94 and 104%.

Table 1. Corrections for determining the percentage of sucrose in sugar solutions by refractometer when readings are made at temperatures other than 20 C [1]

Temp. °C	Percentage Sucrose										
	0	5	10	15	20	25	30	40	50	60	70
					Subtract from the percentage sucrose						
15	027	029	031	033	034	034	035	037	038	039	040
16	022	024	025	026	027	028	028	030	030	031	032
17	017	018	019	020	021	021	021	022	023	023	024
18	012	013	013	014	014	014	014	015	015	016	016
19	016	006	006	07	007	007	007	008	008	008	008
					Add to the percentage sucrose						
21	006	007	007	007	007	008	008	008	008	008	008
22	013	013	014	014	015	015	015	015	016	016	016
23	019	020	021	022	022	023	023	023	024	024	024
24	026	027	028	029	030	030	031	031	031	032	032
25	033	035	036	037	038	038	039	040	040	040	040
26	040	042	043	044	045	046	047	048	048	048	048
27	048	050	052	053	054	055	055	056	056	056	056
28	056	057	060	061	063	063	063	064	064	064	064
29	064	066	068	069	071	072	072	073	073	073	073
30	072	074	077	078	079	080	080	081	081	081	081

Table 2. Table of specific gravities and refractive indices at 20 C of solutions of sucrose [5].

Sucrose % m/m	Specific gravity at 20/20 C	Refractive index n^{20}_D	Sucrose % m/m	Specific gravity at 20/20 C	Refractive index n^{20}_D
0	100000	133299	51	123727	142219
1	100389	133443	52	124284	142432
2	100779	133588	53	124844	142646
3	101172	133733	54	125408	142862
4	101567	133880	55	125976	143080
5	101965	134027	56	126548	143299
6	102366	134176	57	127123	143520
7	102770	134326	58	127703	143742
8	103176	134477	59	128286	143966
9	103586	134629	60	128873	144192
10	103998	134783	61	129464	144420
11	104413	134937	62	130059	144649
12	104831	134093	63	130657	144879
13	105252	132250	64	131260	145112
14	105677	135408	65	131866	145346
15	106104	135567	66	132476	145581
16	106534	134728	67	133090	145819
17	106968	134890	68	133708	146058
18	107404	136053	69	134330	146299
19	107844	136218	70	134956	146541
20	109287	136384	71	135585	146786
21	108733	136551	72	136218	147032
22	109183	136719	73	136856	147279
23	109636	136888	74	137496	147529
24	110092	137059	75	138141	147780
25	110551	13723	76	138790	148033
26	111014	13740	77	139442	148288
27	111480	13758	78	140090	148544
28	111949	13775	79	140758	148803
29	112422	13793	80	141421	149063
30	112898	13811	81	142088	149325
31	113378	13829	82	142759	149589
32	113861	13847	83	143434	149854
33	114347	1385	84	144112	150121
34	114837	13883	85	144794	150391
35	115331	13902	86	145480	
36	115828	13920	87	146170	
37	116329	13939	88	146862	
38	116833	13958	89	147559	
39	117341	13978	90	148259	
40	117853	13997	91	148963	
41	118368	14016	92	149671	
42	118887	14036	93	150381	
43	119410	14056	94	151096	
44	119936	14076	95	151814	
45	120467	14096	96	152535	
46	121001	14117	97	153260	
47	121538	14137	98	153988	
48	122080	14158	99	154719	
49	122625	14179	100	155454	
50	123174	142008			

Table 3. Relation of Brix (Balling) and Baumé [6]

Brix	Baumé	Brix	Baumé
1	0.56	38	20.89
2	1.12	39	21.43
3	1.68	40	21.97
4	2.24	41	22.50
5	2.79	42	23.04
6	3.35	43	23.57
7	3.91	44	24.10
8	4.46	45	24.63
9	5.02	46	25.17
10	5.57	47	25.70
11	6.13	48	26.23
12	6.68	49	26.75
13	7.24	50	27.28
14	7.79	51	27.81
15	8.34	52	28.33
16	8.89	53	28.86
17	9.45	54	29.38
18	10.00	55	29.90
19	10.55	56	30,42
20	11.10	57	30.94
21	11.65	58	31.46
22	12.20	59	31.97
23	12.74	60	32.49
24	13.29	61	33.00
25	13.84	62	33.51
26	14.39	63	34.02
27	14.93	64	34.53
28	15.48	65	35.04
29	16.02	66	35.55
30	16.57	67	36.05
31	17.11	68	36.55
32	17.65	69	37.06
33	18.19	70	37.56
34	18.73	71	38.06
35	19.28	72	38.55
36	19.81	73	39.05
37	20.35	74	39.54

Table 4. Acid Corrections for Obtaining Brix from refractometer reading [7]. (based on citric acid content of citrus juices or other acid-containing sugar solutions). Correction to be added to the Refractometer Reading

% Acid	Corr.	% Acid	Corr.	% Acid	Corr.	% Acid	Corr.
0.0	0.00	7.0	1.34	14.0	2.64	2.10	3.88
0.2	0.04	7.2	1.38	14.2	2.69	21.2	3.91
0.4	0.08	7.4	1.42	14.4	2.72	21.4	3.95
0.6	0.12	7.6	1.46	14.6	2.75	21.6	3.99
0.8	0.16	7.8	1.50	14.8	2.78	21.8	4.02
1.0	0.20	8.0	1.54	15.0	2.81	22.0	4.05
1.2	0.24	8.2	1.58	15.2	2.85	22.2	4.09
1.4	0.28	8.4	1.62	15.4	2.89	22.4	4.13
1.6	0.32	8.6	1.66	15.6	2.93	22.6	4.17
1.8	0.36	8.8	1.69	15.8	2.97	22.8	4.20
2.0	0.39	9.0	1.72	16.0	3.00	23.0	4.24
2.2	0.43	9.2	1.76	16.2	3.03	23.2	4.27
2.4	0.47	9.4	1.80	16.4	3.06	23.4	4.30
2.6	0.51	9.6	1.83	16.6	3.09	23.6	4.34
2.8	0.54	9.8	1.87	16.8	3.13	23.8	4.38
3.0	0.58	10.0	1.91	17.0	3.17	24.0	4.41
3.2	0.62	10.2	1.95	17.2	3.21	24.2	4.44
3.4	0.66	10.4	1.99	17.4	3.24	24.4	4.48
3.6	0.70	10.6	2.03	17.6	3.27	24.6	4.51
3.8	0.72	10.8	2.06	17.8	3.31	24.8	4.54
4.0	0.78	11.0	2.10	18.0	3.35	25.0	4.58
4.2	0.81	11.2	2.14	18.2	3.38	25.2	4.62
4.4	0.85	11.4	2.18	18.4	3.42	25.4	4.66
4.6	0.89	11.6	2.21	18.6	3.46	25.6	4.69
4.8	0.93	11.8	2.24	18.8	3.49	25.8	4.73
5.0	0.97	12.0	2.27	19.0	3.53	26.0	4.76
5.2	1.01	12.2	2.31	19.2	3.56	26.2	4.79
5.4	1.04	12.4	2.35	19.4	3.59	26.4	4.83
5.6	1.07	12.6	2.39	19.6	3.63	26.6	4.86
5.8	1.11	12.8	2.42	19.8	3.68	26.8	4.90
6.0	1.15	13.0	2.46	20.0	3.70	27.0	4.94
6.2	1.19	13.2	2.50	20.2	3.73	27.2	4.97
6.4	1.23	13.4	2.54	20.4	3.77	27.4	5.00
6.6	1.27	13.6	2.57	20.6	3.80	27.6	5.03
6.8	1.30	13.8	2.61	20.8	3.84	27.8	5.06

Table 5. Anthocyanins of common fruits, molecular weights (MW), molar absorbance (ε), wavelength of maximum absorption (λ) [31]

Pigment	MW[*]	LOGε	ε	Solvent	λ
PGD-3-glu	433.2	4.35	22,400	1%HCl/H$_2$O	520
(callistephin)		4.50	31,600	1%HCl/MeOH	516
Cyd-3-gal	445.2	4.48	30,200	1%HCl/MeOH	530
(Idaein)		4.62	41,700	HCl/ETOH	535
		4.49	30,900	HCl/ETOH	535
Cyd-3-rut	595.2	4.46	28,800	1%HCl/H$_2$O	541
Cyd-3-glu	445.2	4.47	29,600	0.01%HCl/MeOH	528
(Chrysanthemin or Asterin)		4.43	26,900	Aqueous Buffer pH 1	510
Dpd-3-glu	465.2	2.90	795	1%HCl/MeOH	543
(Myrtillin)					
Cyd-3-soph	611.2				
Cyd-3-(2^G–xylrut)	727.2				
Cyd-3-(2^G–glurut)	757.2				
Mvd-3-glu	493.5	4.44	28,00	10^{-1}N HCl	520
(Oenin)					
Mvd-3,5-diglu	655.5	4.57	37,700	10^{-1}N HCl	520
(Malvin)					

[*] Molecular weights do not include the chlorine ion or water crystallization

3 Physical Measurements

Definition and Terminology [47-49]: The colour of tropical fruits and fruit products is undoubtedly the first and the most important part of its quality attribute. It is often regarded as an index of general quality determination.

The colour may be observed as the light transmitted through a solution of substances extracted from the food (plant pigments such as carotene, lycopene, chlorophyll and anthocyanins) in a transparent medium by using spectrophotometer or colorimeter. The colour can be also measured as the light reflection from the surface of the food, and will be discussed in this part. Colour which can be judged by the eye as a part of the sensoric analysis will be explained later (chapter 4).

When a photoelectric cell replaces the eye, thus largely eliminating the error due the personal characteristics of each observer, the instrument is termed a photoelectric colorimeter.

Recent advances in the physics and engineering of colour measurement have adequaltely solved the problem of colour measurement of food. These methods, however, are expensive and the specification of the colour does not in itself indicate consumer preference for that item.

The visible region for the human eye is the wavelength between 400 and 750 nm. The mixture of all colours of different wavelength in the visible region is known as white light. The visual colour is complementary to the colour absorbed, that is it is the colour sensation produced by all of the wavelength

Table 6. Relation between absorption and visible colours

Wavelength absorbed nm	Colour absorbed	Visible colour
400–435	violet	yellow-green
435–480	blue	yellow
480–490	green-blue	orange
490–500	blue-green	red
500–560	green	purple
560–580	yellow-green	violet
580–595	yellow	blue
595–605	orange	green-blue
605–750	red	blue-green

Table 7. Physical and sensory terms used to denote different colour attributes

Physical measurment	Sensory term equivalent
Radiant energy	Light
Reflectance	Lightness, Value
Dominant wavelength	Hue, Colour
Purity	Chroma, Saturation, Intensity
Directional reflectance	Gloss, Sheen

minus the wavelengths absorbed. Table 6 gives the relation between absorption and visual colour (Table 6).

A complete specification of colour requires measurement of three recognizable attributes of colour (Table 7):

1. *Hue*, the kind of colour, red, blue or green, at the dominant wavelength.
2. *Saturation or Chroma*, the strength intensity of the colour, vivid colour, dull colour.
3. *Lightness or Value,* when all radiant energy in the visible spectrum is reflected the object appears to be white, if the absorption is complete, the result is a black object. Lightness may be understood as the extent to which the hue is diluted with black.

Equipment: Coloured objects, such as paint, chips, ceramic tiles, etc. are still used as colour standards. An example is the PANTONE colour formula guide 747XR. Due to the uncontrollable fading of pigments, vehicle discolouration and paper aging, the book should be replaced regularly to maintain accurate colour communication.

In order to avoid these difficulties and place the measurement and specification of colour on a scientific basis, The International Commission on Illumination (CIE = Commission Internationale de l'Eclairage, 1931) adopted a set of standards which has made it possible to define the colour in absolute terms.

According to the CIE-system, any colour can be matched exactly by a suitable mixture of only three colours selected from the red or amber (x), green (y) and blue (z) parts of the spectrum (Figs 8 and 9). If the power of the red, green and

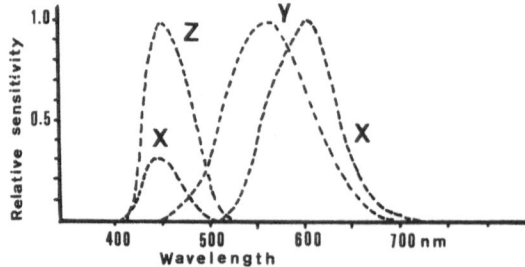

Fig. 8. CIE-System of colours, Z = blue, Y = green and X = red

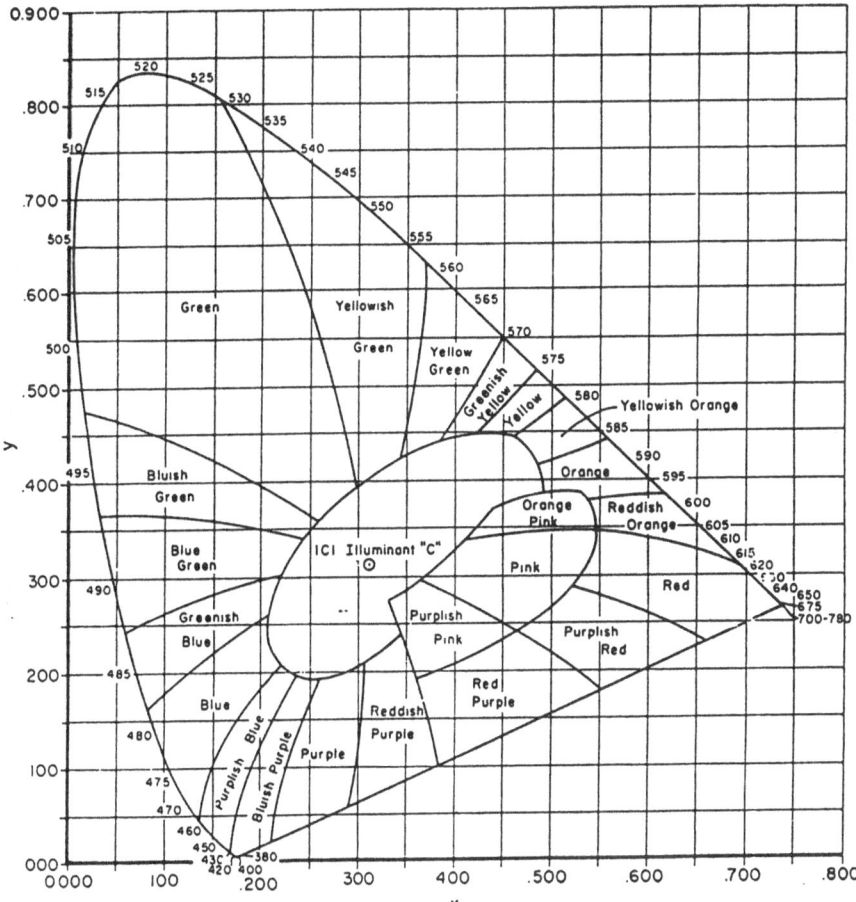

Fig. 9. The (x.y)-chromaticity diagram of the CIE-system

blue lights is suitably chosen, the effect of the combined colours at the point 0 in the middle of the triangle will be white light (Fig. 9). Nearly every colour can be matched at some point in the triangle by a suitable combination of the three colours.

Hunter, with the National Bureau of Standard, US Department of Commerce, in 1952, developed the most successful tricolourimetric system for measuring food colours. Tristimulus amber (red), green and blue filters together with selected photocells and metering circuits provide an approximation of the x, y, and z function of the CIE-system.

In the Hunter-system, the chromaticity plane defined by dimensions a and b and by L (Fig. 9).

- Hunter positive a values indicate redness, and negative a values greenness.
- Hunter positive b values indicate yellowness, and negative b values blueness.
- Hunter L is the visual lightness, 0 = black and 100 = white

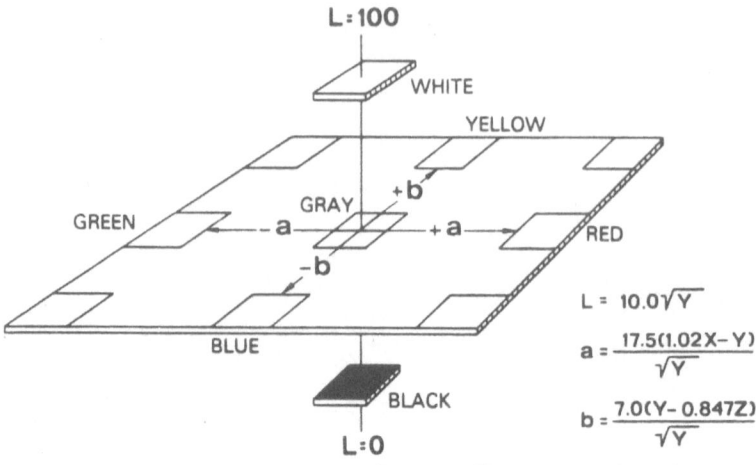

$$L = 10.0\sqrt{Y}$$

$$a = \frac{17.5(1.02X - Y)}{\sqrt{Y}}$$

$$b = \frac{7.0(Y - 0.847Z)}{\sqrt{Y}}$$

Courtesy Hunter Associates Lab. Inc.

Fig. 10. Diagram showing the dimensions of the colour and colour-difference meter, L, a and b colour solid

The a values are functions of x and y, and the b values those of z and y. Hunter values may also be converted to values of the CIE-system.

For example, the colour of mango, passion fruit and kiwi pulp can be expressed numerically using such equipment. It would be

Mango	Passion fruit	Kiwi
L = 45.3	42.3	43.9
a = +29.7	+ 6.4	−12.2
b = +24.8	+30.7	+20.7

To get an idea, what does these numbers mean, please look at Fig. 10.

The colour changes of mango pulp during storage can also be measured and expressed numerically. To express the same colour changes using the sensoric analysis is more difficult and vague.

3.2 Measurement of Consistency

3.2.1 Viscosity

Definition and Terminology [50-53]: Viscosity and texture are properties of appearance with great importance for fruits and their products. Texture measurement of fresh fruits is one of the attributes usually used for maturity determina-

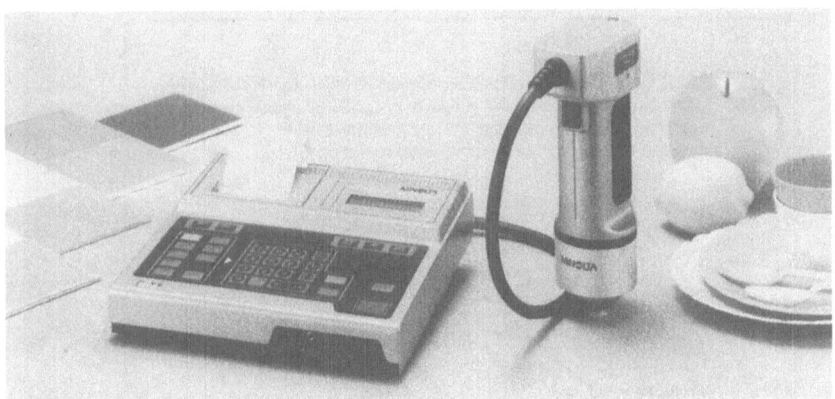

Fig. 11. Minolta Chroma Meter

tion. Viscosity of fruit juices, pulps, jellies and jams have a relation to quality and to processing technology, such as pumping, filling and concentration.

Viscosity or consistency is the friction resulting from the resistance to flow between the liquid layers or the resistance offered by a substance to deformation when subject to a shearing force.

The physical law governing the viscosity was defined by Sir Isaak Newton (1642-1727). Today, fluids may be classified as "Newtonian" or "Non-Newtonian" according to their behaviour at constant temperature under imposed shearing forces (Fig. 11).

Newtonian fluids are chemically "pure" and physically homogeneous. They have a constant consistency value if the static pressure and temperature are fixed. For such substances the consistency is called "viscosity" or "absolute viscosity". When a Newtonian liquid is subjected to shearing stress, a proportional increase in the rate of shear is observed as the stress force is increased (Fig. 11).

Clear juices (apple, grape, cherry and pear juice) and the serums of orange juice and tomato juice (without pulp) are Newtonian fluids.

Non-Newtonian fluids are materials whose resistance to flow changes with a change in rate of shear (Fig. 12.) There are three main types of flow: pseudoplastic, plastic and dilatant fluids: In "plastic fluids" (or Bingham plastics) a definite initial shearing force is required to disrupt the "three-dimensional structure" before any noticeable flow is observed. Once this barrier is overcome, the material follows the flow behaviour of a Newtonian fluid. Tomato ketchup is sometimes stated as being a Bingham plastic. It is because of its "yield value" that it may not start flowing freely from the bottle. After the bottle is struck, however, the minimal pressure required to start flow (yield value) is exceeded, and the ketchup will then pour.

In "pseudoplastic materials" the apparent viscosity decreases as the rate of shear at which the material is tested increases (Fig. 12). Most of the tropical fruit juices, cloudy apple juice, orange juice, tomato juice and concentrates are pseudoplastic fluids.

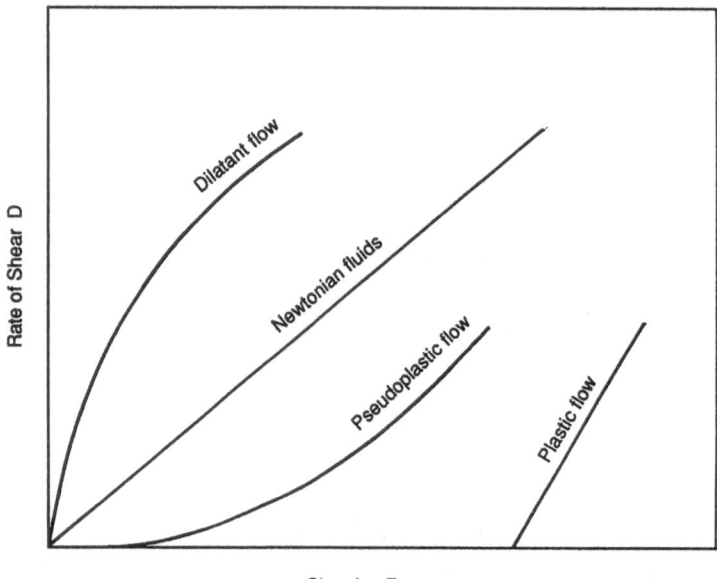

Shearing Force τ

Fig. 12. Fluid flow curves of Newtonian an Non-Newtonian fluids

"Dilatants" are considered to be shear thicking materials. They exhibit an increase in apparent viscosity (thickening) as the rate of shear increases (Fig. 12). Most liquids of this nature return to their original consistency as soon as agitation stop. Examplex of dilatant materials are heavy starch suspensions and some honey types.

Other descriptive terms of consistency frequently encountered are "thixotropy" and "rheopexy". "Thixotropic fluids" are gels, that break up on being shaken and reset on standing. In these fluids, the shear stress (shear force) decreases with the time of shear at a constant rate of shear. At any given time a thixotropic fluid can be regarded as a pseudoplastic. Apple pulp concentrate can be regarded as a thixotropic fluid.

In "rheopectic fluids" the shear stress increases with the time of shear at a constant rate of shear and may be under the same type of system that causes dilatancy. At any particular time, a rheopectic fluid can be regarded as a dilatant. Freshly prepared gelatin gels often exhibit rheopexy.

Thixotropic and rheopectic fluids are time-dependent Non-Newtonian fluids. Plastic, pseudoplastic and dilatant fluids are time-independent Non-Newtonian fluids.

The unit of absolute viscosity is the "poise" after the French physicist Poiseuille (1799-1869). A material requiring a shearing force of one "dyne" per sqare centimeter to produce a rate of shear of one inverse second has a viscosity of one poise. The dyne is a unit of force in the EGS-system of units (Centimeter-Gram-Second). It is the amount of force required to give a mass of 1 gram an acceleration of 1 centimeter per second per second. In the equation f = m a, which is one way of stating Newton's second law of motion, the force f is in dynes when the mass m is given in grams and the acceleration a in cm/sec^2 (centimeters per second per second).

The "centipoise" is one-hundredth of a poise. The absolute viscosity of water at 20 C = 0.01005 poise = 1.0050 centipoise and at 100 C = 0.002838 poise = 0.28380 centipoise.

The international unit system (Systéme International d'Unités = SI) expresses the viscosity (rho) in "Pascal-seconds" (Pa.s) or "millie-Pascal-seconds" (mPa.s). One Pascal-second is equal to ten poise; one millie-Pascal-second is equal to one centipoise.

The kinematic viscosity is the dynamic viscosity divided by the density

ny = rho/s

The kinematic viscosity is expressed in Stokes and centiStokes, 1 Stokes = 1 St = ny, after the British physicist Sir George Stokes (1819-1903).

The previous history of the sample can significantly affect the viscosity measurement. Thus, storage condition and sample preparation techniques, composition and additives, homogeneity of the sample must be designed to minimize their effect on subsequent viscosity tests. Measuring conditions, such as temperature, time, shear rate, aging are the most obvious factors that can have an effect on the rheological behaviour of material. Consider also the fact that most of the fruit juices and concentrates will undergo changes in viscosity during the process of chemical and microbiological reactions. They have also the tendency to separate into non-homogeneous layers.

Viscosity is, nevertheless, more easily measured than some of the properties that affect it, making it a valuable tool for material characterization.

Measurement of Viscositiy [50-53]: Viscosity is the resistance to flow, due to internal friction when fluids are in motion. When a fluid flows over the flat surface of a solid, that layer of the fluid in contact with the solid remains at rest because of adhesion. The adjacent layer of the fluid particles moves slowly over the first layer, the third over the second, and so forth, the speed increasing with the distance from the stationary surface. The fluid is thus subjected to a shearing stress. The viscosity of liquids decreases with the rise of temperature. Apparent viscosity is a measure of resistance to shear or flow at a given rate shear expressed in absolute units. The measurement can be also expressed in time (flow time in seconds) or in maximum distance of flow (in cm).

Many instruments can be used for measuring the viscosity of fruit juices and concentrates, but the recommended types are:

– Capillary tube viscometer
– Rotational viscometer, and
– Distance consistometer.

Capillary Tube Viscometer: These are capillary tubes, which measure the viscosity by determining the *time* required for a given volume to pass a graduation mark under standard condition (e.g. constant-temperature water-bath). By selecting the proper diameter and length of capillary and in some instruments the size of bulb, the viscosity of juices can be measured. The instruments are made of glass, are relatively inexpensive and simple to operate.

Capillary viscometers exist as "absolute" instruments having a known instrument constant and which enable direct calculation of viscosities from flow rate and pressure differential. They can also be supplied as "relative" instruments in which case the user calibrates the instrument with liquids of known viscosity.

The main types of capillary viscometers are:

- Ostwald type viscometer
- Cannon-Fenske type viscometer
- Ubbelohde type viscometer
- IFU type viscometer
- AOAC type viscometer

The Ostwald type viscometer is the main instrument used for the determination of viscosity of fruit juices and for the determination of molecular weight of pectin. It is available in range of capillary diameters to cover a wide range of viscosities (Fig. 13 and Table 8). The Ostwald viscometer is a gravity type where the driving pressure is obtained from the head of the liquid. The instrument measures the kinematic viscosity, since the liquid density affects the flow rate. Kinematic viscosity is the dynamic viscosity divided by density and expressed in centiStokes.

The Cannon-Fenske viscometer (Fig. 13) eliminates the error due to deviation from the vertical. The two bulbs lie exactly on the same axis and the capillary is in line. The disadvantage of having to work with a known volume of liquid in a U-tube viscometer is overcome in the suspended level instruments such as the

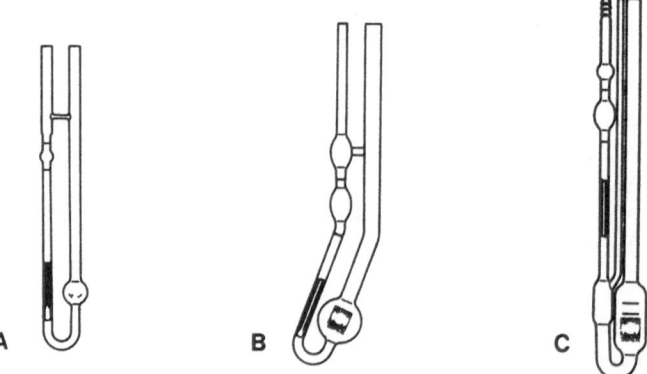

Fig. 13. Ostwald viscometer (*A*), Cannon-Fenske viscometer (*B*) and Ubbelohde Viscometer (*C*)

Table 8. Specification of the Oswald viscometer

Capillary Diam. mm	Flow time of water approx. s	Constant K approx.	Suitable from cSt
0.3	256	0.0043	0.3
0.4	80	0.014	1.0
0.5	34	0.033	2.5
0.6	16	0.069	5.5
0.7	9	0.13	10.0

Ubbelohde viscometer. A third tube joined to the bulb at the lower end of the capillary automatically fixes the lower liquid level. The total activity of commercial pectolytic preparations can be determined by monitoring the decrease in the viscosity of the pectin solution as a function of the reaction time; 9.5 ml of a 0.5% (W/V) solution of high methoxyl pectin (DE 89%) in 0.01 M sodium tartrate buffer pH 3.6 is pipetted into the Ubbelohde glass capillary viscometer (Fig. 13) at 30 C. The flow time of the buffer at 30 C is about 27.3 sec. Enzyme solution (0.5 ml) is injected into the viscometer and mixed by passing air bubbles through it before measuring the flow time. The specific viscosity (rho) is calculated according to the following equation:

$$sp = \frac{t - t_0}{t_0}$$

where:

t_0 = flow time of buffer
t = flow time of the reaction mixture

The reciprocal of specific viscosity (1 : sp) is plotted against the reaction time. The amount of enzyme which reduces the specific viscosity by 50% is calculated and the total activity is determined from the slope of the straight line.

$$(\frac{1}{sp} \text{ vs time})$$

For the determination of apparent viscosity of fruit juices and concentrates the International Federation of Fruit Juice Producers [30] recommends a simple glass tube (Fig. 14) joined to a nozzle through a 5 cm long elastic pipe with a pinch tab. The nozzle is 25 mm long and available in range of capillary diameters (from 6 to 2 mm i.d.) to cover a wide range of viscosities. The rate of flow of a given volume is measured at 200 C. The result includes the time required in s, the nozzle diameter and water value.

The Association of Official Analytical Chemists [8] recommended a capillary viscometer applicable to fruit nectars and fruit juice products (Fig. 15). *Calibration:* Scribe calibration line around outside of reservoir at level reached by water in 13 s under the following conditions. Add water to tube at 24 ± 2 and establish steady flow. Stop flow by placing finger over end of capillary tube. Completely

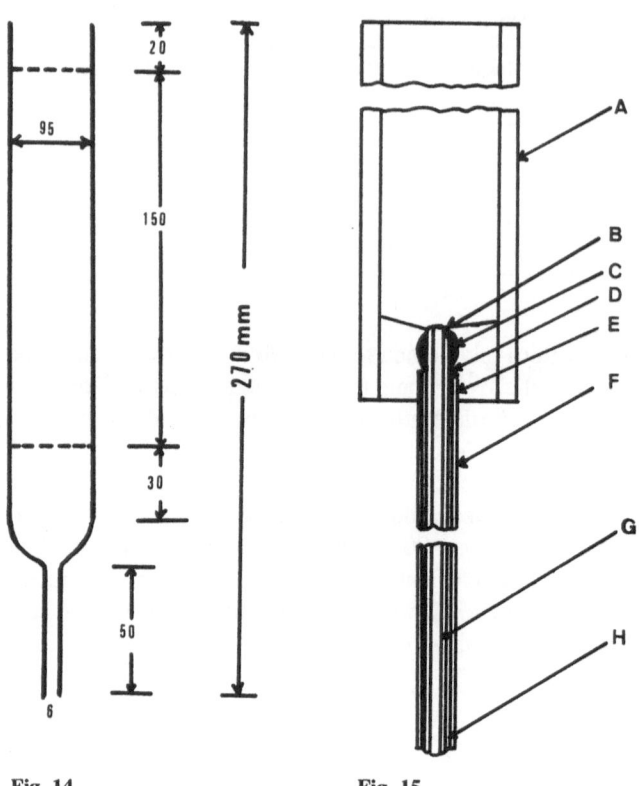

Fig. 14   Fig. 15

Fig. 14. IFU glass tube for the determination of viscosity of fruit juices

Fig. 15. AOAC-capillary viscometer. A = Lucite tube chamber; B = Inner tube, ground 120° included angle; C = Lucite plug, 60° included angle; D = Tygon packing gland, turned 60° included angle (both ends); E = Brass outer tube, ground 60° for packing gland nut; F = Brass tube, chrome plated; G = Inner tube, precision Pyrex glass, i.d. 3 ± 0.001 mm; H = Tygon sleeve

fill tube to overflow point and level off with spatula or by sighting across top of tube. Remove finger from tube and immediately begin timing. Time required for top meniscus to reach calibration line must be 13.0 ± 0.2 sec.

Determination: Clean and dry the apparatus and maintain at 24 ± 0.5 during determination. Adjust sample to 24 ± 0.5 and mix without incorporating air bubbles. Add sample to tube and let flow until steady flow is obtained. Place finger over end of capillary tube to stop flow. Fill tube almost full and check for air bubbles; if air bubbles occur, remove by gently stirring with stirring rod or thermometer (check temperature at this point). Fill to overflow point and level off as in calibration. Remove finger from tube and immediately begin timing. Record time to the nearest 0.1 sec for top of meniscus to reach calibration line.

Rotational viscometer: The design of these instruments is based on the fact that viscosity is the measure of the internal friction of a fluid. This friction becomes apparent when a layer of fluid is made to move in relation to another layer. The greater the friction, the greater the amount of force required to cause this movement, which is called shear. Shearing occurs whenever the fluid is physically moved. Highly viscous fluids, therefore, require more force than less viscous materials. The viscosity is consequently proportional to the shear stress or resistance to flow at a defined speed or shear rate.

Rotational viscometers usually consist of a cylinder rotating in a static measuring cup filled with the sample. The rotor is driven at fixed or programmable speeds by a DC motor utilizing a feed back loop for very accurate speed control. The resistance of the sample to flow causes a very small movement, mounted between the motor and the drive shaft. This movement or deflection is detected by an electronic transducer and indicated by a pointer.

In some instruments the cup containing the sample is rotated. In the liquid is the bob representing a coaxial cylinder. The bob is suspended on a torsion wire and torque resulting from the viscous drag on the bob results in rotation of the graduated scale past a stationary pointer. In some instruments the bob is rotated and the torque on the cup measured, or a constant torque applied and the resulting velocity measured.

Rotational viscometers are more popular than capillary viscometers, because they are more accurate, versatile, simple to operate and can be used for viscosity measurements of Non-Newtonian fluids.

Good experience has been made with viscosimeters of Brookfield, Rotovisco RV 20, Viscotester VT 500 (Haake) and Viscotron (Brabender).

Brookfield viscometers have received wide acceptance in fruit processing plants and in quality control laboratories. Several models are available. All measure viscosity by sensing the torque required to rotate a spindle at constant speed while immersed in the sample fluids. The torque is proportional to the viscous drag on the immersed spindle, and thus to the viscosity of the fluid. The instruments have several outstanding advantages: versatile, each Brookfield viscometer model can measure a wide range of viscosities; multiple speeds and selection of the specific spindle result in a large selection of viscosity for optimum sensitivity and accuracy. The continuous rotation of the spindle allows uninterrupted measurements to be made over long periods of time, permitting the analysis of time-dependent fluid properties. The rate of shear the sample fluid is subjected to is constant, so the instrument is suitable for measuring Newtonian and Non-Newtonian fluids. By rotating the immersed spindle at several different speeds, shear dependent behaviour of Non-Newtonian fluids can be detected and analyzed. Accurate, reproducible viscosity determination can be made quickly without requiring a high degree of operator skill.

Brookfield viscometers are available in two basic types: dial-reading (analog) and digital. The most significant difference between them is the manner in which the viscosity reading is displayed. The dial-reading type is read by noting the position of a pointer in relation to a rotating dial; the digital type is read by means of a 3-digit LED display. In addition, the digital viscometer includes a 0–10 mv output that may be connected to a variety of devices, such as remote displays, controllers, and recorders.

For tropical fruit processing plants the Brookfield dial viscometer (Fig. 16) can be used. It the least expensive Brookfield viscometer and it is suitable for most applications where samples are to be tested over a short period of time and a permanent detailed record of rheological behaviour is not required. Its calibration spring is unaffected by long use of extreme environmental conditions. Reading from the viscometer dial can be directly converted into centipoise units.

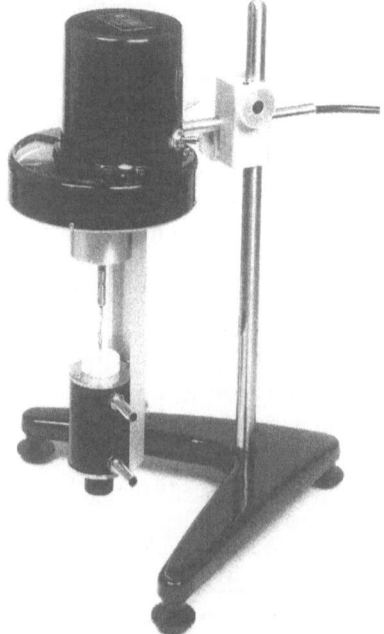

Fig. 16. Brookfield dial reading viscometer

HAAKE Rotovisco RV 20: This instrument has received wide acceptance in the measurement of viscosity of fruit juices and concentrates. The rotovisco RV 20 offers flexibility and has the following advantages besides the features of regular rotational viscometer; viscosity can be measured in an extreme range from 0.02 to 10^9 mPa.s, all flow properties can be determined and measurements take little time and cleaning of sensors is simple.

The basic system of Rotovisco is a complete rotational viscometer for manual operation. Thirty different fixed shear rates from 0.01% to 100% can be present at the rotovisco RV 20 within 3 ranges of the maximum shear rate: 1%, 10% and 100%. The 10 shear rates per range are set in logarithmic steps to provide more data in the interesting low shear range rather than in the high shear range. The preset shear rate (D) is digitally displayed considering the range and step position to avoid calculations or looking up data tables.

The measuring result is shown when the display selector is set to (τ) shearing stress. The shear stress is digitally displayed as a percentage of the maximum achievable value. To calculate viscosity, the shear stress (τ) value has simply to be divided by the shear rate (D) value and multiplied with the sensor factor.

The basic Rotovisco RV 20 is preferably used to determine the flow behaviour of Newtonian, pseudoplastic, plastic and dilatant fluids or to measure the changes of viscosity as a function of temperature.

The actual test temperature is displayed in C when the display selector is in the position (T). The advantage of using one switch is that the rheological parameters: shear stress (τ), shear rate (D) and temperature (T) can be monitored at any time during a test.

For low shear and low viscosity measurements the Rotovisco RV 20–CV 100 is recommended. In the measuring system of the CV 100 the outer cylinder is driven by a DC motor. The inner cylinder is supported by an extremely low friction air bearing used to center the measuring cylinder even for viscoelastic liquids showing normal stress. The movement of the inner cylinder, due to viscosity, is measured by a photodetector which monitors the slightest deflection and activates a motor to compensate for it.

A fully-automatic Rotovisco RV 20 with HAAKE-Rheocontroller RC 20, personal computer and software is also available.

Haake Viscotester VT 500: The VT 500 is a portable, hand-held viscometer for mobile use and can be operated without problems by even untrained lab personnel. The calculated viscosity, shear rate and temperature values can be read directly on the digital display. The speed settings and sensors are compatible with the HAAKE units RV 20.

The display ranges are:

– Viscosity0.5–10^7 mPa.s
– Shear rate4–6000 s^{-1}
– Temperature-30–200 C

Brabender Viscotron: Brabender viscotron is used for routine measurements of viscosity in works laboratories. For highly scientifically measurements the universal Brabender rheometer Rheotron is recommendable. The measuring system of the Viscotron is comparable with the Rotovisco.

3.3 Distance Consistometer

The widely used consistometer for plastic like products, such as tomato paste, jams, jellies, baby foods and fruit purees is the *Bostwick consistometer* (Fig. 17).

The equipment measure the *distance* (cm) over which the material flows on a level surface under its own weight during a given time (30 s). The length of flow is proportional to consistency. The equipment was originally designed by E.P. Bostwick of the U.S. Dept. of Agriculture in 1938 and is officially used by the USDA in establishing the score points for tomato paste and ketchup.

The equipment is made of stainless steel and equipped with a spirit level and two leveling screws. A spring-operated gate is held by a positive release mechanism permitting instantaneous flow of the sample. It is graduated in 0.5 cm divisions to permit accurate measurement of flow. The measurement is made at 20 C and the sample has to be kept several hours at the same temperature.

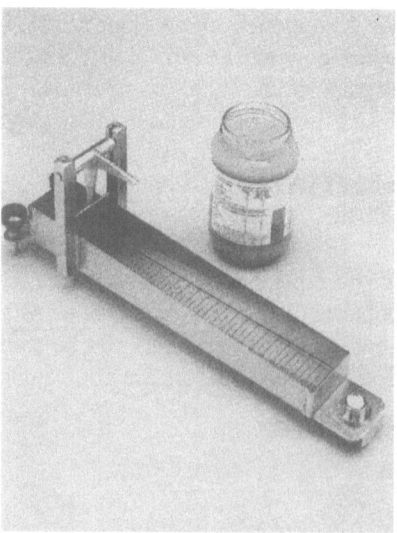

Fig. 17. Bostwick consistometer

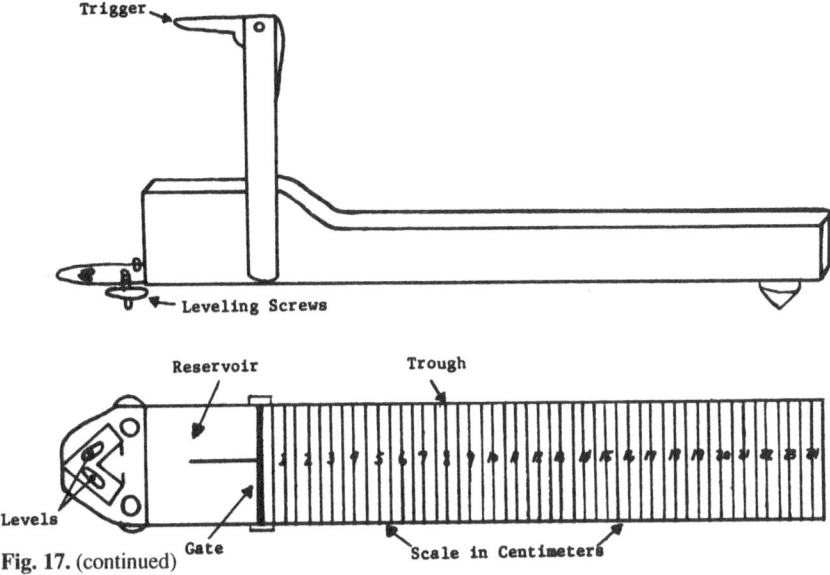

Fig. 17. (continued)

3.3.1 Texture

Definition and Terminology [54-61]: Although there is as yet no single universally accepted definition of food texture, the various proposals are becoming more similar. Texture is difficult to define since it means different things to various people. Generally, *texture* is a group of physical properties that derive from the structure of the food, are sensed by the feeling of touch, are related to the deformation, disintegration and flow of the food under a force. The food texture is a consequence of microstructure, which in turn may be affected by both chemical composition and physical forces. Most of the sensing of food texture occurs in the mouth or between the fingers.

On the other hand, *viscosity* is the internal friction of a fluid or its tendency to resist flow. Texture applies to "solid" foods and viscosity applies to "fluid" foods, but it is well known that the distinction between solids and liquids is not clear cut.

Texture studies are now firmly based on established principles of rheology and the measurement of mechanical properties of foods, cover all the basic aspects of the science of rheology. *Rheology* is the branch of physics dealing with forces and deformations, their relationships and their interrelationssships with time. It is simply the study of deformation and flow of matter, and rheological properties are a major factor in the evaluation of food quality by the sense of touch. The study of food rheology and food texture is a trial to express "mouth-feel" in rheological terms.

Mouth-feel is the mingled experience deriving from the sensation experienced by the lining of the mouth during the ingestion of a food. It is related to viscosity, texture, surface tension, chemical composition ....

The term "*body*" in fruit products is related to consistency, compactness of texture and fullness. It is the textural property producing the mouth-feel sensation of the fruit product.

For those who want to know more about the rheology and texture of food several references are available [54-61].

Texture Measurements Using Commercially Available Instruments: Force measuring instruments are the most common equipments for objective texture measurement. Force has the dimensions mass x length x time^{-2}. The standard unit of force is the Newton (N). A great range of equipment is readily availabe to measure force, deformation and time.

Fruits and fruit products are viscoelastic. The instruments which can be used in a fruit processing plant have the following basic principles:

1. Puncture testing,
2. compression-extrusion, and or
3. shear testing.

3.4 Puncture Testing

Puncture testers are the most widely used instruments in the fruit processing plants because of their simplicity. The puncture test measures the maximum force required to push a punch or probe into food to a given depth at a given weight and time. There are three types of puncture and penetration testers:

– Hand-operated fruit firmness testers,
– Mechanical and motorized puncture testers, and
– Distance measuring instruments.

Hand-Operated Fruit Firmness Testers: These instruments known as fruit "pressure testers" have been widely adopted for field tests to determine the maturity of fruits. They are widely used since the test machine requirements are relatively simple, they are portable, hand-operated and inexpensive (from 30 to 100 US $).

These testers are derived from the improved type of pressure tester developed in 1925 by Magness and Taylor. The test itself is characterized by a force measuring instrument, penetration of the probe in the food causing irreversible crushing, and the depth of penetration is usually held constant. The tester use a spring to measure applied force and the maximum test force is indicated by a spring scale (e.g. a spring housed in a hypodermic syringe).

A number of manufacturers supply hand-held fruit pressure testers, e.g. Chatillon pressure tester and Effi-Gi fruit pressure tester.

Table 9. Hand-operated puncture testers

Type	Force ranges x graduations	Punch Diam. mm	Plunger travel face	Weight g to full scale force cm
Chatillon 719–5	2.2 kg x 50 g	11/8	flat	10180
Chatillon 719–10	4.5 kg x 50 g	11/8	flat	10420
Chatillon 719–20	9.0 kg x 100 g	11/8	flat	10450
Chatillon 719–40	18.0 kg x 200 g	11/8	flat	10500
Chatillon 516–500	500 g x 5 g	0.7–1.6	flat	10180
Chatillon 516–1000	1000 g x 10 g	0.7–1.6	flat	10180
Effi-Gi FT 327	12kg x 250g	11/8	rounded	2170
Effi-Gi FT 011	5 kg x 100 g	11/8	rounded	2170

Chatillon makes two series of testers. The 179 series covers force ranges of 2.2 kg to 18 kg and provides 11 and 8 mm diam. punches (Table 9). The Chatillon 516 series are smaller and lighter instruments (Table 9), with force ranges of 500 and 1000 g. A small chuck at the end of the shaft is used to hold one of the five punches that range from 0.7 to 1.6 mm. The Chatillon tester costs less than 100 US $.

In general, the 11 mm diam. punches are used on most fruits, because the force required will be less than 13 kg. The 8 mm diam. punches are used on very hard fruits, when the force would exceed 13 kg with the larger punch diameter. The small diameter punches of the Chatillon 516 series are recommended for soft fruits.

In Chatillon hand-operated puncture testers, there is no inscribed line indicating how far the punch should penetrate into the fruit, and there is no splash collar.

The *Effi-Gi* tester is the most compact (the smallest and lightest) and most convenient to handle. It can be carried in a pocket and costs less than 50 US $.

Effi-Gi fruit puncture testers are available in two models: Model FT 327 is designed for apples and pears and model FT 011 for citrus, plums etc. The instrument has a dial force gauge and punches having 11 and 8 mm diameter. The face is rounded. The tester has an inscribed line back from the front end of the punch, indicating the depth to which it should be pressed into the sample. A splash collar prevents juice from running back along the shaft.

The spring of both instruments should be calibrated regularly to ensure that they are giving the correct force reading. The companies recalibrate their instruments at the factory for a nominal fee. In Effi-Gi, the zero is adjusted by adding or removing shims to the inside of the instrument. The Chatillon pressure testers have a knob at the end of barrel nearest the pressure tip. Rotating this knob adjusts the zero point. The zero reading should be checked while holding the instrument horizontal before each use and adjusted to the correct value.

Both instruments are held against the surface of the fruit and forced into the fruit with steady pressure (smoothly and gently) to obtain the force required for breaking the flesh. This force is recorded on the pressure tester and used to indicate the maturity of the fruit.

For best results, a suitable sample will be composed of 15-20 fruits, 2 measurements have to be taken on each fruit on opposite sides, near the middle point of each side. Removing a 2 cm diameter disc of peel is recommended.

For the use of the Effi-Gi tester, hold the fruit firmly in the left hand, hold the tester between thumb and forefinger of the right hand, push bottom-commanded indicator hand, place the plunger against the fruit and press with increasing strength until the plunger tip has penetrated into the pulp up to the notch. Slow penetration of the plunger is essential – sharp movements and sudden pressure application may impair the measurements.

The use of hand-operated puncture testers is very simple, but they have one major undesirable feature in that the force is applied manually and the rate of application is not controlled. The technique has, therefore, been adapted to mechanical penetration or deformation mechanisms in order to apply force at a controlled rate and in some devices also in conjunction with an electronic recording system. It goes without saying that such multiple measuring instruments are more expensive.

Mechanical and Motorized Puncture Testers

The commercially available instruments having a mechanical and motorized puncture are:

– *Bloom Gelometer*
– *Stevens LFRA Texture Analyzer*
– *Multiple Measuring Instruments* (will be discussed later)

The Bloom Gelometer is based on the lowering of a plunger a predetermined distance into the product being tested, the force (weight) applied to the plunger to drive it against the resistance of the material is a direct measure of the consistency of the material. The Bloom Gelometer was designed in 1925 to measure the strength of gelatins and gelatin jellies. It can be used today for testing pectin, and jams. The principle of the measurement of Bloom strength is to measure the rigidity of a gel produced under carefully prescribed conditions (standard jelly) using a standard jar. The test solution of gelatin (6.67%) is held for 17 hours at 10 C in a standard jar (a bottle of internal diameter 59 ± 1 mm, 85 mm high, having a capacity of 155 ml. A stopper of rubber 43 mm diam. should fit snugly into the neck of the bottle and should be pierced centrally with an air vent (0.5 mm diam.), and using a Bloom type gelometer, adjusted to give a 4 mm depression and to deliver lead shot at a rate of 200 g per 5 seconds, when the hopper contains 800 g of polished No. 12 chilled shot. The plunger has a diameter of 121.7 mm for gelatin tests and 25.5 mm for gelatin desserts. When the plunger has penetrated 4 mm into the jelly, which usually occurs suddenly, an electrical contact shuts off the flow of shot. The shot is weighed and the weight of shot in grams is expressed as the Bloom value of the gel. The Bloom values of gelatin range from 0 to over 300. The Bloom gelometer is not capable of covering the whole range of gelatins from very high to very low gel strengths without either varying the concentration of gelatin or altering the size of the plunger plummet. This has often led to confusion, if the conditions of measurement are not stated.

Medium rigidity gelatins, between 100 and 200 Bloom, are commercially available.

The Bloom Gelometer is 18 x 19 x 63 cm high and weighs about 13 kg.

The Stevens LFRA Texture Analyzer has replaced the Boucher Electronic Jelly Tester, which is no longer manufactured. It is a development of the Leatherhead Food Research Association (LFRA), England. It was designed as a Bloom test device according to the A.O.A.C. standard and the British standard No. 757, beside a number of other tests (texture of jams, fats, liver paste ...).

The instrument has the standard flat-faced straight-sided acrylic punch (12.7 mm diam.) as the Bloom Gelometer. Punches of other diameters and forms (needle, ball, blade) are also available. The speeds are 0.2, 0.5, 1 and 2 mm/s and the penetration distance is adjusted from 1 to 50 mm in 1 mm-steps.

The punch moves downward at the maximum speed until a force of 5 g is registered, when it automatically steps down to the selected speed and travels at this speed for the selected distance. At the end of the strike it returns to its original position at maximum speed. An electronic load cell senses the force and registers on a digital readout, which shows the maximum force obtained in the test.

The instrument has a capacity of 1000 g (10 N) and with graduations of 1 g. It can be adapted to a 100 g (1 N) force capacity and reading to within 0.1 g for very soft products.

The instrument stands about 50 cm high, 25 cm wide and 25 cm deep, and weighs about 12 kg. A recorder is an optional accessory giving force-distance plots of the puncture tests.

3.5 Distance Measuring Instruments

A great range of instruments are readily available, e.g.

- *Ridgelimeter*
- *SUR Penetrometer*
- *Bostwick Consistometer* (see Sect. 3.2.1)
- *Spreadmeter*

Ridgelimeter was developed for determining the gelling power (in US-SAG according to the recommendation of the "Institute of Food Technologists Committee of Pectin Standardization [62]) and the jelly strength of pectin jellies.

"Grade" of pectin (= SAG-degree) means the weight of sugar which one part by weight of pectin will, under suitable conditions (jelly containing 65% of total sugars at pH 3.0), form a satisfactory jelly. The commercial pectin for marmalade and jam production has a gelling power of 150 US-SAG, which means that 1 kg of standardized pectin will turn 150 kg sugar into a standard gel (SS = 65.0%; gel strength = 23.5% SAG). In other words 1 kg of 150 "jelly grade" pectin can set 150 x 100/65 = 230 kg standard jelly.

The unit of measurement which applies to gelling power was given the English name "sag" because a ridgelimeter measures the sag following a given length of time in a standard jelly.

According to the IFT-Committee the pH of the test jelly is a critical item. It should be pH 3.0. The jelly must contain 65% TSS (650 g of Tss in 1000 g of jelly). The ratio of the actual weight of pectin to that of sugar in the jelly is defined as "assumed grade" for the pectin in the jelly. Hence weight of pectin to be used is 650/assumed grade (if the assumed grade is 150, 4.33 g (= 650/150) of pectin is required). Mix the pectin thoroughly with a part of the sugar before preparation of the jelly. The boiling jelly (exactly 95 C) is poured into the standard ridgelimeter glasses (standard depth and volume). After standing 24 h at 25 C, the jelly is carefully tipped out onto a plate glass, that is furnished with the instrument. The pointer of the dial is moved down close to the surface of the jelly. After exactly 2 min the pointer is moved until it just contacts the jelly. The micrometer screw has 32 threads to an inch so that one revolution moves the point 0.03125 in or 1% of the supported height of the jelly (and is equal to 1% SAG). The scale gives the percentages SAG to the nearest 0.1%.

A jelly of "standard firmness" has a SAG of 23.5%. The true grade of the test is obtained from the formula:

True grade = assumed grade (2.0 – %SAG/23.5).

If a more precise calculation is needed, a conversion curve (Fig. 18) may be used. Figure 18 shows the percentage SAG and the ratio true grade/assumed grade (factor). From the SAG measured, the true grade of the test pectin may be ascertained from the figure.

Assuming that a jelly was made at 200 grade and that it showed 26% SAG, it will be seen from Fig. 18 that the true grade is 0.9 of the assumed grade or 8200 x 0.9 = 180.

Sunkist Growers, the producer of the Ridgelimeter, give all the information needed for using the equipment.

The Ridgelimeter which is used in the standardization method for pectin cannot be used to measure the texture of jams with large fruit particles. Also, it cannot be used to measure the texture of the product in the commercial package. It requires that the jelly be poured into a standard Ridgelimeter glass and allowed to gel in this. The Stevens Texture Analyzer and SUR Penetrometer are better

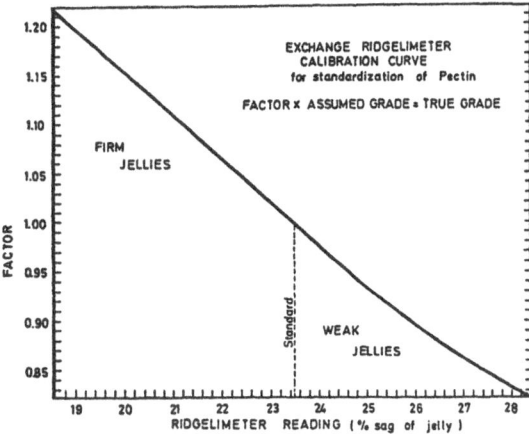

Fig. 18. Relation between ratio „true grades/assumed grades (factor) and per cent SAG (Ridgelimeter, Sunkist Growers)

suited for routine measurement of the texture of jams and jelled products in the commercial package.

Sommer und Runge (SUR) Penetrometer

Penetrometers are made by several manufacturers. The commonly used types are from Precision Scientific and from Sommer und Runge. Both are very similar. The penetration value is defined as the penetration depth (in 0.1 mm), measured at known temperature (25 C), with a penetrating body of known weight and a known penetration time (5 s).

The SUR penetrometer PNR 10 is recommended for both routine quality assurance and for research tasks in a fruit processing plant. It is microprocessor controlled, has a great variation in measurement parameters, has a digital readout and is provided with a limit indicator. The equipment is suitable for testing fruits as well as fruit paste, pulp, jelly, gelatine, jam and marmalade.

The key parts of the SUR penetrometer are an electronic position sensor, which can handle penetrator combination of up to 2 kg weight and a large base plate, which can also support large sample containers with temperature control. The digital display also indicates the upper and lower limits, if such limits are set, in addition to the measured value and test duration. The measurement range of the SUR Penetrometer PNR 10 is from 0 to 48 mm, with a display precision of 0.01 mm in the 0.00 to 10.09 range, and 0.1 in the 10.1 to 48.0 mm range.

For routine quality assurance, a limit indicator is provided for the penetration depth. The test duration can be present digitally beween 0.1 second and 2.78 hours. The automatic default value set when the unit is switched on, is 5 seconds.

The penetrometer has a weight of 10 kg and dimensions of 36 x 30 x 41 ... 52 cm. It is recommended for texture measurement of fruits and its products, by choosing different plungers. For fruits the pin needle penetrator is recommended: plunger 15 g, loading weight 2 g and stainless steel pin needle 3 g. It can be also adapted to measuring the deformability of the fruits. The needle of the penetrometer is replaced by a light weight flat circular disk (e.g. of 102.5 g and a plunger of 47.5 g). For marmalade, jam and jelly a test body of large cross section is recommended: such as hollow aluminium rod of 7 g-10 mm with 3 interchangeable weights 3, 13, and 23 g; or the test cylinders with tip and a plunger 15 g; or a penetration cone of 102.5 g and a plunger of 47.5 g.

Spreadmeter

The spreadmeter is simply an acrylic plate with 19 concentric circles with constant separation drawn on the underside. When determining the fluidity of the sample a suitable amount (e.g. the content of a 450 g jar) is placed in the middle of the plate. After 5 minutes the average "flowing out" of the sample is measured. The test is very simple and useful (Fig. 19).

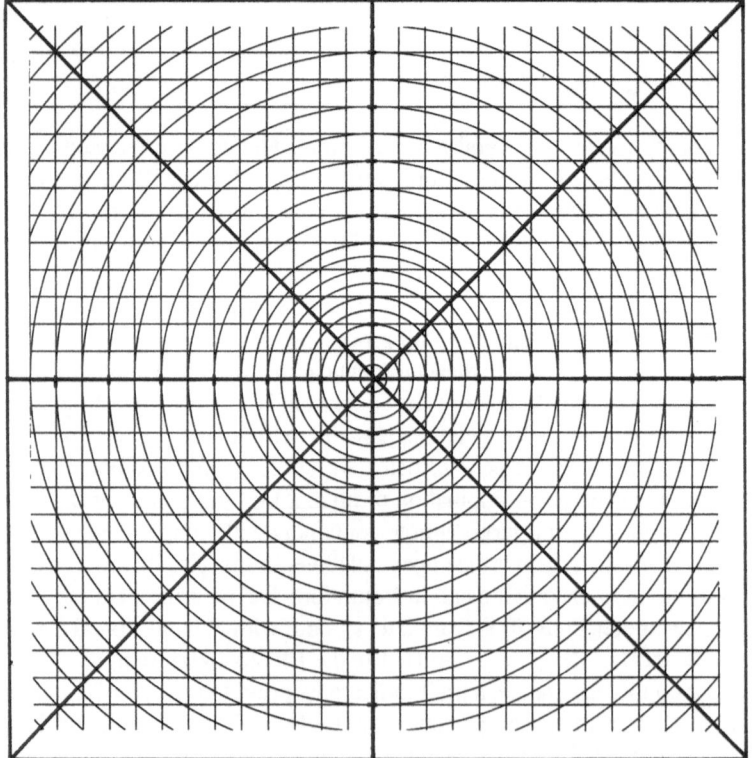

Fig. 19. Spreadmeter

Multiple Measuring Instruments

With such instruments a load/deformation curves are recorded while a plunger is brought down into the food at a pre-set speed. By choosing different plungers and test cells, various texture properties can be determined (such as shear, puncture, compression, energy of deformation ...). The measured variables are force (F), time (T) and distance (D). The recording system plots the complete history of force changes throughout the test, so-called "Texture Profile Analysis" (TPA). This means much more information than the one point maximum force reading that is obtained from simple instruments. The multiple measuring instruments are suited for research purposes and cost over 10 000 US $. The following instruments are recommended:

- *INSTRON Universal Testing Instrument*
- *Texture Test System (Kramer Shear Press)*
- *General Foods Texturometer*

3.6 Measurement of Water Activity

Definition and Terminology [63-66]: The growth and metabolism of microorganisms invariably depend on the presence of water in an available form. The water activity (a_w) of a food provides an objective measure of the amount of water available for microbiological usage. Microbial growth does not occur below certain a_w values, which vary according to the specific microorganism:

Bacteria	
Escherichia coli	0.95
Lactobacillus viridescens	0.95
Bacillus cereus	0.95
Salmonella spp.	0.95
Yeast	
Saccharomyces cerevisiae	0.90
S. bailii	0.80
S. rouxii	0.62
Mould	
Aspergillus flavus	0.78
A. candidus	0.75
Eurotium (Aspergillus) amstelodami	0.70
E. echinulatum	0.62

Generally, bacteria require an a_w above 0.91, yeast above 0.88 and molds above 0.75. Osmotolerant yeasts and some molds can grow above 0.60.

Water activity (a_w) is a measure of the relative volatility or escaping potential of water in a food. It is defined as the quotient of the existing water vapour pressure and the maximum possible water vapour pressure at a given temperature.

It is defined as:

$a_w = Pi/Po$

where

Pi = vapour pressure of water in equilibrium with food
Po = vapour pressure of water at the same temperature.

Water activity and equilibrium relative humidity (ERH) are related by :

$aw = \% ERH/100$

The movement of water between a food product and its environment is governed by the water activity (a_w). For example, if the a_w of a food is greater than that of the atmosphere, water will be lost from the food to the air resulting in dehydration of the food, until a_w of the air and the food become the same or when a dynamic equilibrium condition is reached. The activity of water in foods may be affected by dissolved solutes, capillary forces, and interaction with macromolecules.

The Water Sorption Isotherm is a plot of moisture content as a function of a_w at constant temperature. It can be obtained by the adsorption process of exposing a dry food to increasing a_w environment.

Moisture sorption isotherms provide a means of predicting the storage stability of a food product based on its moisture content. The fact that the sorption isotherms of dried fruits lie above those of cereal grains indicates that the former adsorb more water at the same a_w. This explaines why dried fruits remain stable

Table 10. Water activity of some intermediate moisture foods ($a_w = 0.60$–0.90 water content 10–40 % and the relation to spoilage and deterioration [66]

a_w	Food	Spoilage and Deterioration
1.00	water rich food	
0.90	fruit juice concentrates	min. for bacterial growth
0.85	jams, marmalade	min. for yeast growth
0.80	soft dried fruits (figs)	
0.75	Honey, soft dried fruits	min. for mould growth
0.70	dried fruits	
0.65	dried fruits	max. for Maillard-Reaction
0.60	confectionery	min. for osmophilic yeast and mould
0.50	dry food	
0.40		
0.30		
0.20		
0.10		
0.00		max. for oxidation

Table 11. Water activity of water at different temperatures [66]

C	a_w
0	1.00
−5	0.95
−20	0.82
−30	0.75

at a moisture content of 22-24%, while most cereal grains spoil at 18% moisture content. In general, the more humectants (e.g. sugar, salt) a food contains, the lower its a_w at a certain moisture content.

Equipment

The most common techniques for measuring the a_w include:

– electric hygrometry,
– vapour pressure manometry,
– isopiestic equilibration methods,
– phychromery and hair hygrometry.

A number of *electric hygrometers* are available commercially. Electric hygrometers consist of a potentiometer, a sample/sensor holder and a sensor. Most of these instruments utilize sensors with a coated hygroscopic salt (e.g. LiCl) or thin polymer film. Water activity is determined indirectly from the change in electrical conductance or capacitance as the sensor absorbs moisture from the sample.

The instruments read out or record either directly in a_w units or ERH. The accuracy at 25 C in the 0.75 to 0.99 a_w range is usually better than 0.005 a_w.

Two common makes of electric hygrometers available are:
Rotronic Hygroskop D
Novasina Hygrometer DAL 02
The other three techniques mentioned are rather out of date.

4 Sensory Analysis

In the absence of direct methods of measuring taste and flavour, sensory evaluation of fresh fruits and final products provides a practical and rapid test for quality. The procedures are neither laborious nor time-consuming, yet the data obtained will enable those evaluating foods to judge the quality of the product and the consumer preference without following detailed chemical and microbiological methods usually employed in evaluating foodstuffs.

Sensory analysis determines mainly the following quality attributes: Appearance, including colour, size, shape, uniformity and absence of defects.

Taste and flavour, including texture, consistency, viscosity, feeling, taste and smell (odour). Any of all of the senses may be used in such tests, namely: Smell (olfactory-mucous membrane of the nose, olfactory, epithelium), taste (gustatory-mucous membrane of the tongue, palate, and throat), sight (visual-eyes), hearing (auditory, aural-ears) and touch (haptic-tactile nerves in general). The detectors for sensory impressions are the nerve cells. These work differently in different individuals. The results will be influenced by many factors, e.g. experiences, eating habits, mental and psychological attitudes....

When people are used as the measuring instruments (colour, flavour, taste), the importance of planning, control and standardization of the sensory analyses cannot be over emphasized. Sensory analysis endeavors to overcome this difficulty by good *organization of the tests* and by the use of *statistical test designs*. It is important to become familiar with statistical analysis before using sensory methods of analysis (esp. difference tests: paired, triangle, duo-trio-test, and ranking tests).

The organization of the test includes:

- *Sampling*, random and representative sampling.
- *Preparation of test material*, all samples to be evaluated should be coded. They should never be suggestive to the panel judges. The order of presentation should also be randomized within each test session.
- *Definition of question required to be answered*, the questionnaire or the score card should be prepared carefully for each test. It should be simple and clear. The card should be clearly typed or printed.

– *Selection of the appropriate statistical method*, the statistical design to decide the quality of the sample and the statistical treatment of the data should be selected carefully. These will depend on the defined objective of the test. The method should be appropriate to the problem and the situation. Table 12 gives the most common sensory test and associated standards. The statistical design is used mainly to minimize the error and to measure variables or to establish the significance of results.

– *Selection of test subjects*, usually a small number (5–10) of selected panels are principally used in routine quality control and in research and development studies. A greater number (more than 100) chosen at random to represent a cross-section of the population for which the product is intended can be used in some studies.

The selected panel judges should be interested in such activity, willing to give conscientious decisions and above all be cooperative.

It goes without saying that the they should not have a cold or nasal or digestion difficulties. They should have good health, average sensitivity, a high degree of personal integrity, intellectual curiosity, a feeling of responsibility, ability to express sensations in words, memory, ability to concentrate, ability to make judgements and to avoid bias. Availability is also an important factor.

– *Laboratory set up and equipment*, a simple lab with an atmosphere of relaxed concentration, temperature at about 20 C and relative humidity of 70–80%. The lab should be clean, quiet, odour-free, uniformly illuminated, equipped with extra coloured lights (e.g. yellow) whenever needed to mask small differences in the colour of the samples. The lab should be preferably air-conditioned and free from any distraction. A pale green colour for walls has proved to be very calming for the eyes.

Table 12. Sensory evalution methods and associated standards

Method	Test	DIN	ISO
Difference tests	Paired comparison	10954/1977	5495/1983
(qualitative)	Triangle test	10951/1978	4120/1983
	Duo-Trio		
Rating tests	Ranking	10963/1982	8587/1988
(quantitative	Points scoring	10952/1984	
differences)			
Descriptive tests	Descriptive profile	10964/1983	
Sensitivity	Threshold	10959/1977	3591/1977

Sensory test procedure:

– The test should be held between 9–11 am or 2–4 pm
– There should be no set time limit for an individual's judgement.
– Judges should be provided with water, so that they may rinse their mouths between samples if they so desire.

- All directions to panel members should be in writing and should be very clear (typed or printed).
- All the scoring, remarks and recommendations of the judges should be in writing.
- The judges are separated from each other. Judging should be done in individual booths.
- No communication between panel members should be allowed except for consultation with the panel leader.
- Odour testing by sniffing should be done before tasting.
- In sample presentation colour differences have to be masked by coloured light or by using coloured beakers or colouring the samples.

4.2 Statistical Test Designs

The statistical design and the statistical evaluation of the results are may be the most important points in the sensory analysis. Sensory analysis without using suitable statistical methods has no value. The commonly used tests are described in Table 12. Several references are available [67–73].

4.2.1 Difference Tests

The commonly used quality difference tests are: pair comparison, triangle test and duo-trio test. These are designed to determine whether a difference occurs between the flavour (or colour or taste) of *two* samples. As differences are very small, these tests are quite difficult and require a high degree of concentration. Managers often cannot concentrate enough because they are too busy and are therefore not suitable as permanent panel members. In order to avoid any bias, the samples must have the same appearance, shape, temperature, consistency, and equal amounts have to be presented.

Paired Comparison Test

This test is designed to compare between two samples for specific characteristics. The test requires 5–10 trained or 70–80 untrained panelists. Several pairs are given to each panel member. Each pair consists of two unknown samples, one the standard or control and other an experimental one. The panelists are asked to

- indicate if the samples are different and/or

– select the sample having the greater or lesser degree of intensity of a specified sensory characteristic.

If more than two samples (two treatments) are being considered, each treatment is compared with every other in the series. The number of sample pairs presented at one session is limited by the degree of fatigue induced by sensory testing.

In the paired comparison test, the chance probability of placing the sample in a certain order is one-half. The test is used for new product development, product improvement, quality maintenance and also for consumer preference (but in consumer analysis, the judgement asked for is for overall quality).

The Binomial distribution(Table 13. Sect. 4.4) is used for the statistical analysis of the data.

Triangle Test

The most common sensory test used in fruit processing plants and quality control laboratories is the triangle test. The test is utilized when evaluating a small number of samples and when attempting to determine small differences in flavour, colour and taste between the samples. It is used also for training and testing panels.

There are two types of triangle test:

– The simple triangle test (one task) and
– the extended triangle test (additional tasks).

Three samples are presented to the judges, two alike and one different. The judges are asked to identify the odd sample. A positive answer is required even it is a question. Since all three samples are unknown, the chance probability of placing the sample in a certain order is one-third.

Two samples A and B can be presented in two combinations AAB and BBA and for replication in six different arrangements:

ABB BAB BBA BAA ABA AAB

The Binomial distribution (Table 14) is used for the statistical analysis of the data obtained by the triangle test. The test requires 5–10 trained panelists.

Two of the three samples are identical, determine the odd sample and indicate the degree of difference.

Set No.	Code no. of odd sample	the best sample	Score	Comment on the samples
I				
II				
III				
IV				
V				

Score
0 = no difference between the samples
1 = small difference
2 = noticeable difference
3 = great difference

It is also possible to study the degree of difference in the triangle test (see the evaluation card for triangle test).

In the extended triangle test, additional tasks are to be fulfilled: defining the degree of difference or characterizing the difference. Only the extended triangle test gives valuable information.

For the statistical analysis of the obtained data use the Tables 13–15 [74], Sect. 4.4.

Triangular-Scoring Flavour Panel

This panel is conducted in the same manner as the triangle test panel, however, the judges are asked to indicate their preferences for the samples on a numerical scoring scale. Only those preferences scores where the judges could identify the like samples are analyzed for significance (only results of panel members with correct differentiation can be analyzed further).

When, for example, 11 people out of 20 correctly analyzed the odd sample, the result is "significant", with 13 correct answers "highly significant" and with 14 correct answers "very highly significant". In the statistical analysis of the extended triangle test, table of significance (Table 15) have to be consulted in addition.

Duo-Trio Test

The test is so called because it is intermediate between the duo (paired) and trio (triangle) test.

This test employs three samples: two identical and one different. The panel is first given one of the pair of identical samples as known reference sample R and then other two successively in random order, and asked to match one of these with the first. A positive answer is required even if it is a guess. The chance probability of placing the sample in a certain order is one-half.

The significance at various probability levels for duo-trio test is given in Table 15, Sect. 4. 4.

The directions for the panelists:

– The first sample "R" given is the reference sample.
– Taste it carefully.
– From the pair of coded samples next given, judge which sample is the same as "R".

– A positive answer is to be made even if it is a guess.

The duo-trio test is used for detecting differences and also for training and testing panels.

With the known control in the duo-trio-test less retesting is necessary, therefore it is used for samples with after-taste or off-notes.

For the statistical analysis of the data obtained use the Tables 13–15 [74], see Sect. 4. 4.

4.2.2 Quantitative Difference Tests

Such tests are used for determining order according to one or more specific characteristic or determining preference used in product and process improvement, selection of the best sample, consumer preference analysis and also the selection of a trained panel. The tests require 5–10 trained panelists or 70–80 untrained.

The common quantitative difference tests used are: Ranking, numerical scoring and composite scoring tests.

Ranking Test

In this test, judges are asked to rank samples (3–6) in decreasing or increasing order on the basis of a single characteristic. A control need not be identified. All the samples should be presented simultaneously and must be coded. The trained panelists are asked to rank all samples according to the intensity of the specified characteristic.

Note that ranking tests have made it clear that the human senses cannot differentiate quantities as well as qualities. Describe first the character note, then add the intensity. Ask always first: what do I taste/smell and secondly "how strongly do I perceive the single notes."

In consumer analysis, the untrained panelists are asked to rank the coded samples according to their preference.

The rank analysis (Tables 16 and 17, see Sect. 4.4) are used for the statistical analysis of data [75].

A triangle test panel can then be conducted to find out if the differences between the best and the poorest (or second sample) samples are significant.

Judges, who decide that samples are very good or very bad are usually mercurial and extrovert types and tend to use the ends of the scale – to them samples are either superb or execrable. Both these extreme types of charcter are best avoided.

Please rank the samples in numerical order according to your preference or intensity of taste (or flavour or colour...).

Intensity/Preference		(score)	sample code no.	comments
First	(excellent)	(100)		
Second	(good)	(75)		
Third	(fair)	(50)		
Fourth	(poor)	(25)		

comments = type of off-taste (off-flavour, colour changes etc).

Numerical Scoring Test

In this test, the samples are presented to each panelist in random order. The panelists are asked to score the samples on a specific scale for a particular characteristic indicating the rating of the samples. The panelists are trained to follow the sensory characteristics descriptions and scores. Without this understanding the rating will not be of any use. The simplest scale for statistical analysis and for routine evaluation is a numerical scaling from 1 to 10, with being:

10	excellent or perfect,
9, 8, 7	good,
6, 5, 4	fair,
3, 2	poor, and
1	off or unacceptable

Several factors can be scored on the same sheet (Colour, flavour, viscosity, taste...).

In the *composite scoring test* the rating scale is defined so that specific characteristics of the product are rated separately. The definition of the rating scale is weighted so that the most important characteristics will account for a large part of the total score. The resulting scores are compounded for any one panelist to arrive at a composite score. The panelist are trained to evaluate the dimensions of the individual quality characteristics critically, and in the use of the weighted scale.

The test is recommended for the quality control of fruit products. It is helpful in grading products and comparison of quality attributes by indicating which characteristic is at fault in a poor product. The composite scoring test is recommended for new product development and for quality maintenance during production and marketing.

The statistical analysis of the data is according to the analysis of variance Tables 18, 19, 20 and 21 [76–77], see Sect. 4. 4.

Quality	Score	I	II	samples III	IV	V
colour	20					
odour	30					
taste	50					
total	100					

Comments to sample

I   :
II   :
III :
IV   :
V   :

Please score the samples according to the following descriptions:

quality description	possible score for colour	odour	taste
Excellent	20	30	50
Good	15	20	40
Fair	10	15	30
Poor	5	10	15

This card is recommended for fruit juices and nectars.

4.2.3 Descriptive Flavour Profile

This test is used in new product development and storage studies. About 5 trained people (specially in the quality control technique) are required. The sample characteristics is expressed in common terms and the intensity is expressed on agreed scale.

4.2.4 Threshold Test

Threshold test is a sensitivity test used mainly for determination of the threshold of flavour compounds (pure substances). It is also recommended for selecting panel members for evaluating ingredients (esp. food additives), packaging material and for the detection of off-flavour in fruit products.

Threshold is defined as the minimum concentration of a substance at which a transition or sensation in a series or judgements occurs (at which a transition occurs from no sensation to sensation).

For selecting panels a series of samples with increasing concentrations of the four taste qualities (sweet, salty, sour, bitter) is used. The panelists are asked to describe the taste or to give intensity scores.

The test can be also used for the determination of the threshold of benzoic acid in fruit juices (the minimum concentration which can be detected by panelists).

Evaluation Card for Threshold Test

(for the determination of threshold or for the selection and training of the panelists).

Date: _____

Name: _____

You receive a series of samples with increasing concentrations of one of the four taste qualities (salty, sweet, sour, bitter). Start with sample 1 and continue with 2, 3 etc, Retasting of already tasted solutions is not allowed. Describe the taste or give intensity scores.

Sample No.	Describtion of taste	Intensity score	Comments
I			
II			
III			
IV			
V			
VI			
VII			
VIII			
IX			
X			

Intensity scale
0 = No taste (water taste)
1 = Different from water, but taste quality not identifiable
2 = Threshold very weak
3 = Weak
4 = Medium
5 = Strong
6 = Very strong

For the preparation of threshold test solution according to Jellinek [68] see Tables 22–25.

4.3 Selection and Training of Panel Members

The panel is not composed of experts in the ordinary sense, as in wine, coffee and tea testing. The real expert is always of value, the problem is when and how to use him.

In a fruit processing plant, there are only a few people available to take part in the panel test. Therefore, a selected group of trained judges is required.

The panel members, ususally five to ten, can be recruited from the available technical personnel, and to qualify for membership they must be congenial, normally perceptive and interested in the work. An older person may be less sensitive but may concentrate better and have experience. Men and women, smokers and non-smokers are in general equally sensitive to food. Only when smokers stop smoking, do they become more sensitive to spoilage odors and flavours. It was at one time assumed that non-smokers made better judges than smokers, but there is very little evidence to support this assumption.

Those who "dislike" the particular food product under test are undesirable as judges.

Generally, women are more sensitive to sweet and salty, while men are more sensitive to sour. The sensitivity for sweet, often also for sour, decreases with age. People who add a lot of salt to their food, as a food habit in many tropical countries, show only slight sensitivity to salt.Candidates should have healthy teeth, otherwise tasting, especially of sour samples, may cause pain. If different metals are used for dental filling, a metallic taste may be perceived, even when the sample has no metallic off-taste.

Each prospective member is tested for organoleptic sensitivity:

1. The first exercise is the recognition test of the four basic tastes: Sweet, salty, sour and bitter, when given in the form of the following aqueous solution:

   | sweet | 1% | sugar | salty | 0.1% | salt |
   | sour | 0.06% | citric acid | bitter | 0.03% | caffeine. |

2. The second exercise can be made using the threshold test. In the threshold test the sensitivity toward the four basic tastes is measured quantitatively for each panel member. Concentration series with sucrose, sodium chloride, citric acid and caffeine are prepared and presented in order of increasing intensity, Tables 22–25.

   A low recognition threshold with aqueous series does not necessarily correspond with a high sensitivity to foods. Therefore, threshold tests should not play a too important role in a sensory course. It can be regarded as a basic exercise. We do not consume only aqueous solutions in our daily life. In this tests the water quality plays an important role. Neither with distilled, double distilled nor with demineralized water were satisfactory results obtained. It is best to boil water for 10 minutes and then decant it after cooling.

   Preparation of the test solutions (Tables 22–25) has to be done with precision.

Only personnel trained in laboratory work should prepare the solutions. Samples can be coded before the preparation. In case candidates help in sample preparation, coding is better done afterwards. Use of numbers 1, 2, 3, 4 and letters a, b, c, d for coding may cause bias. Therefore, coding with 3-- or 4-- digit number or with 2, 4, 6 or 1, 3, 5, 7 is recommended. In order to avoid bias, the sample amount must be equal in all beakers (30 ml). The samples have to be placed in a row and tested from left to right. The best neutralizing agent for this test is water at 35–40C.

The test sheet must be as simple as possible, in order to save writing and testing time. But it must give the impression of importance and difficulty. Because, we usually do not pay enough attention to apparently easy tasks and therefore do them incorrectly. Retesting is allowed, but do not retest too often. The first impression often is the correct one. With repeated testing, fatigue may lead to errors.When the tests are carried out in a hurry they provide only questionable results.

The panel members should not be too hungry when coming to the test as this might cause stomach trouble. One hour after a light meal is the best time. As a physiological effect it can happen that the low concentrations of a salty and sour series taste sweet and those of a bitter series taste salty.

3. The third exercise is to use of ranking test in ranking the series of sucrose, sodium chloride and citric acid solutions in increasing order of intensity. A test can be made also with colour series presented in a test tube stand. The tubes with the samples can be held against a light background. Later tests should be made with food colours, with various fruit drinks with fruit flavours added to fruit drinks.

4. The fourth exercise is to use the triangle test for training and testing panels. As differences are usually very small, triangle test are quite difficult and require a high degree of training and concentration.

5. The selected candidate can be then interviewed to check motivation, interest, experience and personality.

The screening and training of judges absorbs a lot of time and effort, sometimes this is a problem in a fruit processing plant, and the people differ significantly in their "trainability". From our experience, motivation and interest of panel members are the most important factors.

4.4 Statistical Tables

For the statistical analysis of the results obtained in the sensory evaluation use the Tables 13–21 [74–77].

Table 13. Minimum numbers of agreeing judgments necessary to establish significance at various probability levels for the paired-preference test (two-tailed, p = 1/2)[a] [74]

No. of trials (n)	Probability levels						
	0.05	0.04	0.03	0.02	0.01	0.005	0.001
7	7	7	7	7			
8	8	8	8	8	8		
9	8	8	9	9	9	9	
10	9	9	9	10	10	10	
11	10	10	10	10	11	11	11
12	10	10	11	11	11	12	12
13	11	11	11	12	12	12	13
14	12	12	12	12	13	13	14
15	12	12	13	13	13	14	14
16	13	13	13	14	14	14	15
17	13	14	14	14	15	15	16
18	14	14	15	15	15	16	17
19	15	15	15	15	16	16	17
20	15	16	16	16	17	17	18
21	16	16	16	17	17	18	19
22	17	17	17	17	18	18	19
23	17	17	18	18	19	19	20
24	18	18	18	19	19	20	21
25	18	19	19	19	20	20	21
26	19	19	19	20	20	21	22
27	20	20	20	20	21	22	23
28	20	20	21	21	22	22	23
29	21	21	21	22	22	23	24
30	21	22	22	22	23	24	25
31	22	22	22	23	24	24	25
32	23	23	23	23	24	25	26
33	23	23	24	24	25	25	27
34	24	24	24	25	25	26	27
35	24	25	25	25	26	27	28
36	25	25	25	26	27	27	29
37	25	26	26	26	27	28	29
38	26	26	27	27	28	29	30
39	27	27	27	28	28	29	31
40	27	27	28	28	29	30	31
41	28	28	28	29	30	30	32
42	28	29	29	29	30	31	32
43	29	29	30	30	31	32	33
44	29	30	30	30	31	32	34
45	30	30	31	31	32	33	34
46	31	31	31	32	33	33	35
47	31	31	32	32	33	34	36
48	32	32	32	33	34	35	36
49	32	33	33	34	34	35	37
50	33	33	34	34	35	36	37
60	39	39	39	40	41	42	44
70	44	45	45	46	47	48	50
80	50	50	51	51	52	53	56
90	55	56	56	57	58	59	61
100	61	61	62	63	64	65	67

Table 14. Minimum numbers of correct judgments to establish significance at various probability levels for the triangle test (one-tailed, $p = 1/3$) [74]

No. of trials (n)	Probability levels						
	0.05	0.04	0.03	0.02	0.01	0.005	0.001
5	4	5	5	5	5	5	
6	5	5	5	5	6	6	
7	5	6	6	6	6	7	7
8	6	6	6	6	7	7	8
9	6	7	7	7	7	8	8
10	7	7	7	7	8	8	9
11	7	7	8	8	8	9	10
12	8	8	8	8	9	9	10
13	8	8	9	9	9	10	11
14	9	9	9	9	10	10	11
15	9	9	10	10	10	11	12
16	9	10	10	10	11	11	12
17	10	10	10	11	11	12	13
18	10	11	11	11	12	12	13
19	11	11	11	12	12	13	14
20	11	11	12	12	13	13	14
21	12	12	12	13	13	14	15
22	12	12	13	13	14	14	15
23	12	13	13	13	14	15	16
24	13	13	13	14	15	15	16
25	13	14	14	14	15	16	17
26	14	14	14	15	15	16	17
27	14	14	15	15	16	17	18
28	15	15	15	16	16	17	18
29	15	15	16	16	17	17	19
30	15	16	16	16	17	18	19
31	16	16	16	17	18	18	20
32	16	16	17	17	18	19	20
33	17	17	17	18	18	19	21
34	17	17	18	18	19	20	21
35	17	18	18	19	19	20	22
36	18	18	18	19	20	20	22
37	18	18	19	19	20	21	22
38	19	19	19	20	21	21	23
39	19	19	20	20	21	22	23
40	19	20	20	21	21	22	24
41	20	20	20	21	22	23	24
42	20	20	21	21	22	23	25
43	20	21	21	22	23	24	25
44	21	21	22	22	23	24	26
45	21	22	22	23	24	24	26
46	22	22	22	23	24	25	27
47	22	22	23	23	24	25	27
48	22	23	23	24	25	26	27
49	23	23	24	24	25	26	28
50	23	24	24	25	26	26	28
60	27	27	28	29	30	31	33
70	31	31	32	33	34	35	37
80	35	35	36	36	38	39	41
90	38	39	40	40	42	43	45
100	42	43	43	44	45	47	49

Table 15. Minimum numbers of correct judgments to establish significance at various probability levels for paired-difference and duo trio tests (one-tailed, p = 1/2) [74]

No. of trials (n)	Probability levels						
	0.05	0.04	0.03	0.02	0.01	0.005	0.001
7	7	7	7	7	7		
8	7	7	8	8	8	8	
9	8	8	8	8	9	9	
10	9	9	9	9	10	10	10
11	9	9	10	10	10	11	11
12	10	10	10	10	11	11	12
13	10	11	11	11	12	12	13
14	11	11	11	12	12	13	13
15	12	12	12	12	13	13	14
16	12	12	13	13	14	14	15
17	13	13	13	14	14	15	16
18	13	14	14	14	15	15	16
19	14	14	15	15	15	16	17
20	15	15	15	16	16	17	18
21	15	15	16	16	17	17	18
22	16	16	16	17	17	18	19
23	16	17	17	17	18	19	20
24	17	17	18	18	19	19	20
25	18	18	18	19	19	20	21
26	18	18	19	19	20	20	22
27	19	19	19	20	20	21	22
28	19	20	20	20	21	22	23
29	20	20	21	21	22	22	24
30	20	21	21	22	22	23	24
31	21	21	22	22	23	24	25
32	22	22	22	23	24	24	26
33	22	23	23	23	24	25	26
34	23	23	23	24	25	25	27
35	23	24	24	25	25	26	27
36	24	24	25	25	26	27	28
37	24	25	25	26	26	27	29
38	25	25	26	26	27	28	29
39	26	26	26	27	28	28	30
40	26	27	27	27	28	29	30
41	27	27	27	28	29	30	31
42	27	28	28	29	29	30	32
43	28	28	29	29	30	31	32
44	28	29	29	30	31	31	33
45	29	29	30	30	31	32	34
46	30	30	30	31	32	33	34
47	30	30	31	31	32	33	35
48	31	31	31	32	33	34	36
49	31	32	32	33	34	34	36
50	32	32	33	33	34	35	37
60	37	38	38	39	40	41	43
70	43	43	44	45	46	47	49
80	48	49	49	50	51	52	55
90	54	54	55	56	57	58	61
100	59	60	60	61	63	64	66

Table 16. Rank total required for significance at p = 5% [75]

Number of panel tests	Number of samples in each panel test										
	2	3	4	5	6	7	8	9	10	11	12
2	–	–	–	–	–	–	–	–	–	–	–
	–	–	–	3–9	3–11	3–13	4–14	4–16	4–18	5–19	5–21
3	–	–	–	4–14	4–17	4–20	4–23	5–25	5–28	5–31	5–34
	–	4–8	4–11	5–13	6–15	6–18	7–20	8–22	8–25	9–27	10–29
4	–	5–11	5–15	6–18	6–22	7–25	7–29	8–32	8–36	8–40	9–43
	–	5–11	6–14	7–17	8–20	9–23	10–26	11–29	13–31	14–34	15–37
5	–	6–14	7–18	8–22	9–26	9–31	10–35	11–39	12–43	12–48	13–52
	6–9	7–13	8–17	10–20	11–24	13–27	14–31	15–35	17–38	18–42	20–45
6	7–11	8–16	9–21	10–26	11–31	12–36	13–41	14–46	15–51	17–55	18–60
	7–11	9–15	11–19	12–24	14–28	16–32	18–36	20–40	21–45	23–49	25–53
7	8–13	10–18	11–24	12–30	14–35	15–41	17–46	18–52	19–58	21–63	22–69
	8–13	10–18	13–22	15–27	17–32	19–37	22–41	24–46	26–51	28–56	30–61
8	9–15	11–21	13–27	15–33	17–39	18–46	20–52	22–58	24–64	25–71	27–77
	10–14	12–20	15–25	17–31	20–36	23–41	25–47	28–52	31–57	33–63	36–68
9	11–16	13–23	15–30	17–37	19–44	22–50	24–57	26–64	28–71	30–78	32–85
	11–16	14–22	17–28	20–34	23–40	26–46	29–52	32–58	35–64	38–70	41–76
10	12–18	15–25	17–33	20–40	22–48	25–55	27–63	30–70	32–78	34–86	37–93
	12–18	16–24	19–31	23–37	26–44	30–50	33–57	37–63	40–70	44–76	47–83
11	13–20	16–28	19–36	22–44	25–52	28–60	31–68	34–76	36–85	39–93	42–101
	14–19	18–26	21–34	25–41	29–48	33–55	37–62	41–69	45–76	49–83	53–90
12	15–21	18–30	21–39	25–47	28–56	31–65	34–74	38–82	41–91	44–100	47–109
	15–21	19–29	24–36	28–44	32–52	37–59	41–67	45–75	50–82	54–90	58–98
13	16–23	20–32	24–41	27–51	31–60	35–69	38–79	42–88	45–98	49–107	52–117
	17–22	21–31	26–39	31–47	35–56	40–64	45–72	50–80	54–89	59–97	64–105
14	17–25	22–34	26–44	30–54	34–64	38–74	42–84	46–94	50–104	54–114	57–125
	18–24	23–33	28–42	33–51	38–60	44–68	49–77	54–86	59–95	65–103	70–112
15	19–26	23–37	28–47	32–58	37–68	41–79	46–89	50–100	54–111	58–122	63–132
	19–26	25–35	30–45	36–54	42–63	47–73	53–82	59–91	64–101	70–110	75–120
16	20–28	25–39	30–50	35–61	40–72	45–83	49–95	54–106	59–117	63–129	68–140
	21–27	27–37	33–47	39–57	45–67	51–77	57–87	63–97	69–107	75–117	81–127
17	22–29	27–41	32–53	38–64	43–76	48–88	53–100	58–112	63–124	68–136	73–148
	22–29	28–40	35–50	41–61	48–71	54–82	61–92	67–103	74–113	81–123	87–134
18	23–31	29–43	34–56	40–68	46–80	51–93	57–105	62–118	68–130	73–143	79–155
	24–30	30–42	37–53	44–64	51–75	58–86	65–97	72–108	79–119	86–130	93–141
19	24–33	30–46	37–58	43–71	49–84	55–97	61–110	67–123	73–136	78–150	84–163
	25–32	32–44	39–56	47–67	54–79	62–90	69–102	76–114	84–125	91–137	99–148
20	26–34	32–48	39–61	45–75	52–88	58–102	65–115	71–129	77–143	83–157	90–170
	26–34	34–46	42–58	50–70	57–83	65–95	73–107	81–119	89–131	97–143	105–155

Table 17. Rank total required for significance at p = 1% [75]

Number of panel tests	Number of samples in each panel test										
	2	3	4	5	6	7	8	9	10	11	12
2	–	–	–	–	–	–	–	–	–	–	–
	–	–	–	–	–	–	–	–	3–19	3–21	3–23
3	–	–	–	–	–	–	–	–	4–29	4–32	4–35
	–	–	–	4–14	4–17	4–20	5–22	5–25	6–27	6–30	6–33
4	–	–	–	5–19	5–23	5–27	6–30	6–34	6–38	6–42	7–45
	–	–	5–15	6–18	6–22	7–25	8–28	8–32	9–35	10–38	10–42
5	–	–	6–19	7–23	7–28	8–32	8–37	9–41	9–46	10–50	10–55
	–	6–14	7–18	8–22	9–26	10–30	11–34	12–38	13–42	14–46	15–50
6	–	7–17	8–22	9–27	9–33	10–38	11–43	12–48	13–53	13–59	14–64
	–	8–16	9–21	10–26	12–30	13–35	14–40	16–44	17–49	18–54	20–58
7	–	8–20	10–25	11–31	12–37	13–43	14–49	15–55	16–61	17–67	18–73
	8–13	9–19	11–24	12–30	14–35	16–40	18–49	19–51	21–56	23–61	25–66
8	9–15	10–22	11–29	13–35	14–42	16–48	17–55	19–61	20–68	21–75	23–81
	9–15	11–21	13–27	15–33	17–39	19–45	21–51	23–57	25–63	28–68	30–74
9	10–17	12–24	13–32	15–39	17–46	19–53	21–60	22–68	24–75	26–82	27–90
	10–17	12–24	15–30	17–37	20–43	22–50	25–56	27–63	30–69	32–76	35–82
10	11–19	13–27	15–35	18–42	20–50	22–58	24–66	26–74	28–82	30–90	32–98
	11–19	14–26	17–33	20–40	23–47	25–55	28–62	31–69	34–76	37–83	40–90
11	12–21	15–29	17–38	20–46	22–55	25–63	27–72	30–80	32–89	34–98	37–106
	13–20	16–28	19–36	22–44	25–52	29–59	32–67	35–75	39–82	42–90	45–98
12	14–22	17–31	19–41	22–50	25–59	28–68	31–77	33–87	36–96	39–105	42–114
	14–22	18–30	21–39	25–47	28–56	32–64	36–72	39–81	43–89	47–97	50–106
13	15–24	18–34	21–44	25–53	28–63	31–73	34–83	37–93	40–103	43–113	46–123
	15–24	19–33	23–42	27–51	31–60	35–69	39–78	44–86	48–95	52–104	56–113
14	16–26	20–36	24–46	27–57	31–67	34–78	38–88	41–99	45–109	48–120	51–131
	17–25	21–35	25–45	30–54	34–64	39–73	43–83	48–92	52–102	57–111	61–121
15	18–27	22–38	26–49	30–60	34–71	37–83	41–94	45–105	49–116	53–127	56–139
	18–27	23–37	28–47	32–58	37–68	42–78	47–88	52–98	57–108	62–118	67–128
16	19–29	23–41	28–52	32–64	36–76	41–87	45–99	49–111	53–123	57–135	62–146
	19–29	25–39	30–50	35–61	40–72	46–82	51–93	56–104	61–115	67–125	72–136
17	20–31	25–43	30–55	35–67	39–80	44–92	49–104	53–117	58–129	62–142	67–154
	21–30	26–42	32–53	38–64	43–76	49–87	55–98	60–110	66–121	72–132	78–143
18	22–32	27–45	32–58	37–71	42–84	47–97	52–110	57–123	62–136	67–149	72–162
	22–32	28–44	34–56	40–68	46–80	52–92	59–103	65–115	71–127	77–139	83–151
19	23–34	29–47	34–61	40–74	45–88	50–102	56–115	61–129	67–142	72–156	77–170
	24–33	30–46	36–59	43–71	49–84	56–96	62–109	69–121	76–133	82–146	89–158
20	24–36	30–50	36–64	42–78	48–92	54–106	60–120	65–135	71–149	77–163	82–178
	25–35	32–48	38–62	45–75	52–88	59–101	66–114	73–127	80–140	87–153	94–166

Table 18. Variance ratio-5% points for distribution of F [76]

$n2$ \ $n1$	1	2	3	4	5	6	8	12	24	Inf.
1	161.45	199.50	215.70	224.60	230.16	234.00	238.90	243.91	249.05	254.31
2	18.51	19.00	19.16	19.25	19.30	19.33	19.37	19.41	19.45	19.50
3	10.13	9.55	9.28	9.12	9.01	8.94	8,84	8.74	8.64	8.53
4	7.71	6.94	6.59	6.39	6.26	6.16	6.04	5.91	5.77	5.63
5	6.61	5.79	5.41	5.19	5.05	4.95	4.82	4.68	4.53	4.37
6	5.99	5.14	4.76	4.53	4.39	4.28	4.15	4.00	3.84	3.67
7	5.59	4.74	4.35	4.12	3.97	3.87	3.73	3.57	3.41	3.23
8	5.32	·4.46	4.07	3.84	3.69	3.58	3.44	3.28	3.12	2.93
9	5.12	4.26	3.86	3.63	3.48	3.37	3.23	3.07	2.90	2.71
10	4.96	4.10	3.71	3.48	3.33	3.22	3.07	2.91	2.74	2.54
11	4.84	3.98	3.59	3.36	3.20	3.09	2.95	2.79	2.61	2.40
12	4.75	3.89	3.49	3.26	3.11	3.00	2.85	2.69	2.50	2.30
13	4.67	3.81	3.41	3.18	3.03	2.92	2.77	2.60	2.42	2.21
14	4.60	3.74	3.34	3.11	2.96	2.85	2.70	2.53	2.35	2.13
15	4.45	3.68	3.29	3.06	2.90	2.79	2.64	2.48	2.29	2.07
16	4.49	3.63	3.24	3.01	2.85	2.74	2.59	2.42	2.24	2.01
17	4.45	3.59	3.20	2.96	2.81	2.70	2.55	2.38	2.19	1.96
18	4.41	3.55	3.16	2.93	2.77	2.66	2.51	2.34	2.15	1.92
19	4.38	3.52	3.13	2.90	2.74	2.63	2.48	2.31	2.11	1.88
20	4.35	3.49	3.10	2.87	2.71	2.60	2.45	2.28	2.08	1.84
21	4.32	3.47	3.07	2.84	2.68	2.57	2.42	2.25	2.05	1.81
22	4.30	3.44	3.05	2.82	2.66	2.55	2.40	2.23	2.03	1.78
23	4.28	3.42	3.03	2.80	2.64	2.53	2.38	2.20	2.00	1.76
24	4.26	3.40	3.01	2.78	2.62	2.51	2.36	2.18	1.98	1.73
25	4.24	3.39	2.99	2.76	2.60	2.49	2.34	2.16	1.96	1.71
26	4.23	3.37	2.98	2.74	2.59	2.47	2.32	2.15	1.95	1.69
27	4.21	3.35	2.96	2.73	2.57	2.46	2.31	2.13	1.93	1.67
28	4.20	3.34	2.95	2.71	2.56	2.44	2.29	2.12	1.91	1.65
29	4.18	3.33	2.93	2.70	2.55	2.43	2.28	2.10	1.90	1.64
30	4.17	3.32	2.92	2.69	2.53	2.42	2.27	2.09	1.89	1.62
40	4.08	3.23	2.84	2.61	2.45	2.34	2.18	2.00	1.79	1.51
60	4.00	3.15	2.76	2.53	2.37	2.25	2.10	1.92	1.70	1.39
120	3.92	3.07	2.68	2.45	2.29	2.18	2.02	1.83	1.61	1.25
Inf.	3.84	3.00	2.60	2.37	2.21	2.10	1.94	1.75	1.52	1.00

Table 19. Variance ratio-1% points for distribution of F [76]

n_2 \ n_1	1	2	3	4	5	6	8	12	24	Inf.
1	4052.20	4999.50	5403.40	5624.60	5763.60	5859.00	5981.10	6106.30	6234.60	6365.90
2	98.50	99.00	99.17	99.25	99.30	99.33	99.37	99.42	99.46	99.50
3	34.12	30.82	29.46	28.71	28.24	27.91	27.49	27.05	26.60	26.12
4	21.20	18.00	16.69	15.98	15.52	15.21	14.80	14.37	13.93	13.46
5	16.26	13.27	12.06	11.39	10.97	10.67	10.29	9.89	9.47	9.02
6	13.74	10.93	9.78	9.15	8.75	8.47	8.10	7.72	7.31	6.88
7	12.25	9.55	8.45	7.85	7.46	7.19	6.84	6.47	6.07	5.65
8	11.26	8.65	7.59	7.01	6.63	6.37	6.03	5.67	5.28	4.86
9	10.56	8.02	6.99	6.42	5.80	5.47	5.11	4.73	4.31
10	10.04	7.56	6.55	5.99	5.64	5.39	5.06	4.71	4.33	3.91
11	9.65	7.20	6.22	5.67	5.32	5.07	4.74	4.40	4.02	3.60
12	9.33	6.93	5.95	5.41	5.06	4.82	4.50	4.16	3.78	3.36
13	9.07	6.70	5.74	5.21	4.86	4.62	4.30	3.96	3.59	3.16
14	8.86	6.51	5.56	5.03	4.70	4.46	4.14	3.80	3.43	3.00
15	8.68	6.36	5.42	4.89	4.56	4.32	4.00	3.67	3.29	2.87
16	8.53	6.23	5.29	4.77	4.44	4.20	3.89	3.55	3.18	2.75
17	8.40	6.11	5.18	4.67	4.34	4.10	3.79	3.46	3.08	2.65
18	8.29	6.01	5.09	4.58	4.25	4.01	3.71	3.37	3.00	2.57
19	8.18	5.93	5.01	4.50	4.17	3.94	3.63	3.30	2.92	2.49
20	8.10	5.85	4.94	4.43	4.10	3.87	3.56	3.23	2.86	2.42
21	8.02	5.78	4.87	4.37	4.04	3.81	3.51	3.17	2.80	2.36
22	7.94	5.72	4.82	4.31	3.99	3.76	3.45	3.12	2.75	2.31
23	7.88	5.66	4.76	4.26	3.94	3.71	3.41	3.07	2.70	2.26
24	7.82	5.61	4.72	4.22	3.90	3.67	3.36	3.03	2.66	2.21
25	7.77	5.57	4.68	4.18	3.86	3.63	3.32	2.99	2.62	2.17
26	7.72	5.53	4.64	4.14	3.82	3.59	3.29	2.96	2.58	2.15
27	7.68	5.49	4.60	4.11	3.78	3.56	3.26	2.93	2.55	2.10
28	7.64	5.45	4.57	4.07	3.75	3.53	3.23	2.90	2.52	2.06
29	7.60	5.42	4.54	4.04	3.73	3.50	3.20	2.87	2.49	2.03
30	7.56	5.39	4.51	4.02	3.70	3.47	3.17	2.84	2.47	2.01
40	7.31	5.18	4.31	3.83	3.51	3.29	2.99	2.66	2.29	1.80
60	7.08	4.98	4.13	3.65	3.34	3.12	2.82	2.50	2.12	1.60
120	6.85	4.79	3.95	3.48	3.17	2.96	2.66	2.34	1.95	1.38
Inf.	6.64	4.60	3.78	3.32	3.02	2.80	2.51	2.18	1.79	1.00

Table 20. Multiple F tests-significant studentized ranges for a 5% level multiple range test [77]

n_2 \ p	2	3	4	5	6	7	8	9	10	12	14	16	18	20	50	100
1	18.00	18.00	18.00	18.00	18.00	18.00	18.00	18.00	18.00	18.00	18.00	18.00	18.00	18.00	18.00	18.00
2	6.09	6.09	6.09	6.00	6.00	6.00	6.09	6.09	6.00	6.09	6.09	6.09	6.09	6.09	6.09	6.09
3	4.50	4.50	4.50	4.50	4.50	4.50	4.50	4.50	4.50	4.50	4.50	4.50	4.50	4.50	4.50	4.50
4	3.93	4.01	4.02	4.02	4.02	4.02	4.02	4.02	4.02	4.02	4.02	4.02	4.02	4.02	4.02	4.02
5	3.64	3.74	3.79	3.83	3.83	3.83	3.83	3.83	3.83	3.83	3.83	3.83	3.83	3.83	3.83	3.83
6	3.46	3.58	3.64	3.68	3.68	3.68	3.68	3.68	3.68	3.68	3.68	3.68	3.68	3.68	3.68	3.68
7	3.35	3.47	3.54	3.58	3.60	3.61	3.61	3.61	3.61	3.61	3.61	3.61	3.61	3.61	3.61	3.61
8	3.26	3.39	3.47	3.52	3.55	3.56	3.56	3.56	3.56	3.56	3.56	3.56	3.56	3.56	3.56	3.56
9	3.20	3.34	3.41	3.47	3.50	3.52	3.52	3.52	3.52	3.52	3.52	3.52	3.52	3.52	3.52	3.52
10	3.15	3.30	3.37	3.43	3.46	3.47	3.47	3.47	3.47	3.47	3.47	3.47	3.47	3.48	3.48	3.48
11	3.11	3.27	3.35	3.39	3.43	3.44	3.45	3.46	3.46	3.46	3.46	3.46	3.47	3.48	3.48	3.48
12	3.08	3.23	3.33	3.36	3.40	3.42	3.44	3.44	3.46	3.46	3.46	3.46	3.47	3.48	3.48	3.48
13	3.06	3.21	3.30	3.35	3.38	3.41	4.42	3.44	3.45	3.45	3.46	3.46	3.47	3.47	3.47	3.47
14	3.03	3.18	3.27	3.33	3.37	3.39	3.41	3.42	3.44	3.45	3.46	3.46	3.47	3.47	3.47	3.47
15	3.01	3.16	3.25	3.31	3.36	3.38	3.40	3.42	3.43	3.44	3.45	3.46	3.47	3.47	3.47	3.47
16	3.00	3.15	3.23	3.30	3.34	3.37	3.39	3.41	3.43	3.44	3.45	3.46	3.47	3.47	3.47	3.47
17	2.98	3.13	3.22	3.28	3.33	3.36	3.38	3.40	3.42	3.44	3.45	3.46	3.47	3.47	3.47	3.47
18	2.97	3.12	3.21	3.27	3.32	3.35	3.37	3.39	3.41	3.43	3.45	3.46	3.47	3.47	3.47	3.47
19	2.96	3.11	3.19	3.26	3.31	3.35	3.37	3.39	3.41	3.43	3.44	3.46	3.47	3.47	3.47	3.47
20	2.95	3.10	3.18	3.25	3.30	3.34	3.36	3.38	3.40	3.43	3.44	3.46	3.46	3.47	3.47	3.47
22	2.93	3.08	3.17	3.24	3.29	3.32	3.35	3.37	3.39	3.42	3.44	3.45	3.46	3.47	3.47	3.47
24	2.92	3.07	3.15	3.22	3.28	3.31	3.34	3.37	3.38	3.41	3.44	3.45	3.46	3.47	3.47	3.47
26	2.91	3.06	3.14	3.21	3.27	3.30	3.34	3.36	3.38	3.41	3.43	3.45	3.46	3.47	3.47	3.47
28	2.90	3.04	3.13	3.20	3.26	3.30	3.33	3.35	3.37	3.40	3.43	3.45	3.46	3.47	3.47	3.47
30	2.89	3.04	3.12	3.20	3.25	3.29	3.32	3.35	3.37	3.40	3.43	3.44	3.46	3.47	3.47	3.47
40	2.86	3.01	3.10	3.17	3.22	3.27	3.30	3.33	3.35	3.39	3.42	3.44	3.46	3.47	3.47	3.47
60	2.83	2.98	3.08	3.14	3.20	3.24	3.28	3.31	3.33	3.37	3.40	3.43	3.45	3.47	3.48	3.48
100	2.80	2.95	3.05	3.12	3.18	3.22	3.26	3.29	3.32	3.36	3.40	3.42	3.45	3.47	3.53	3.53
Inf.	2.77	2.92	3.02	3.09	3.15	3.19	3.23	3.26	3.29	3.34	3.38	3.41	3.44	3.47	3.61	3.67

Table 21. Multiple F tests-significant studentized ranges for a 1% level multiple range test [77]

n_2 \ p	2	3	4	5	6	7	8	9	10	12	14	16	18	20	50	100
1	90.00	90.00	90.00	90.00	90.00	90.00	90.00	90.00	90.00	90.00	90.00	90.00	90.00	90.00	90.00	90.00
2	14.00	14.00	14.00	14.00	14.00	14.00	14.00	14.00	14.00	14.00	14.00	14.00	14.00	14.00	14.00	14.00
3	8.26	8.50	8.60	8.70	8.80	8.90	8.90	9.00	9.00	9.00	9.10	9.20	9.30	9.30	9.30	9.30
4	6.51	6.80	6.90	7.00	7.10	7.10	7.20	7.20	7.30	7.30	7.40	7.40	7.50	7.50	7.50	7.50
5	5.70	5.96	6.11	6.18	6.26	6.33	6.40	6.44	6.50	6.60	6.60	6.70	6.70	6.80	6.80	6.80
6	5.24	5.51	5.65	5.73	5.81	5.88	5.95	6.00	6.00	6.10	6.20	6.20	6.30	6.30	6.30	6.30
7	4.95	5.22	5.37	5.45	5.53	5.61	5.69	5.73	5.80	5.80	5.90	5.90	6.00	6.00	6.00	6.00
8	4.74	5.00	5.14	5.23	5.32	5.40	5.47	5.51	5.50	5.60	5.70	5.70	5.80	5.80	5.80	5.80
9	4.60	4.86	4.99	5.08	5.17	5.25	5.32	5.36	5.40	5.50	5.50	5.60	5.70	5.70	5.70	5.70
10	4.48	4.73	4.88	4.96	5.06	5.13	5.20	5.24	5.28	5.36	5.42	5.48	5.54	5.55	5.55	5.55
11	4.39	4.63	4.77	4.86	4.94	5.01	5.06	5.12	5.15	5.24	5.28	5.34	5.38	5.39	5.39	5.39
12	4.32	4.55	4.68	4.76	4.84	4.92	4.96	5.02	5.07	5.13	5.17	5.22	5.24	5.26	5.26	5.26
13	4.26	4.48	4.62	4.69	4.74	4.84	4.88	4.94	4.98	5.04	5.08	5.13	5.14	5.15	5.15	5.15
14	4.21	4.42	4.55	4.63	4.70	4.78	4.83	4.87	4.91	4.96	5.00	5.04	5.06	5.07	5.07	5.07
15	4.17	4.37	4.50	4.58	4.64	4.72	4.77	4.81	4.84	4.90	4.94	4.97	4.99	5.00	5.00	5.00
16	4.13	4.34	4.45	4.54	4.60	4.67	4.72	4.76	4.79	4.84	4.88	4.91	4.93	4.94	4.94	4.94
17	4.10	4.30	4.41	4.50	4.56	4.63	4.68	4.72	4.75	4.80	4.83	4.86	4.88	4.89	4.89	4.89
18	4.07	4.27	4.38	4.46	4.53	4.59	4.64	4.68	4.71	4.76	4.79	4.82	4.84	4.85	4.85	4.85
19	4.05	4.24	4.35	4.43	4.50	4.56	4.61	4.64	4.67	4.72	4.76	4.79	4.81	4.82	4.82	4.82
20	4.02	4.22	4.33	4.40	4.47	4.53	4.58	4.61	4.65	4.69	4.73	4.76	4.78	4.79	4.79	4.79
22	3.99	4.17	4.28	4.36	4.42	4.48	4.53	4.57	4.60	4.65	4.68	4.71	4.74	4.75	4.75	4.75
24	3.96	4.14	4.24	4.33	4.39	4.44	4.49	4.53	4.57	4.62	4.64	4.67	4.70	4.72	4.74	4.74
26	3.93	4.11	4.21	4.30	4.36	4.41	4.46	4.50	4.53	4.58	4.62	4.65	4.67	4.69	4.73	4.73
28	3.91	4.08	4.18	4.28	3.34	4.39	4.43	4.47	4.51	4.56	4.60	4.62	4.65	4.67	4.72	4.72
30	3.89	4.06	4.16	4.22	4.32	4.36	4.41	4.45	4.48	4.54	4.58	4.61	4.63	4.65	4.71	4.71
40	3.82	3.99	4.10	4.17	4.24	4.30	4.34	4.37	4.41	4.46	4.51	4.54	4.57	4.59	4.69	4.69
60	3.76	3.92	4.03	4.12	4.17	4.23	4.27	4.31	4.34	4.39	4.44	4.47	4.50	4.53	4.66	4.66
100	3.71	3.86	3.98	4.06	4.11	4.17	4.21	4.25	4.29	4.35	4.38	4.42	4.45	4.48	4.64	4.65
Inf.	3.64	3.80	3.90	3.98	4.04	4.09	4.14	4.17	4.20	4.26	4.31	4.34	4.38	4.41	4.60	4.68

Table 22. Preparation of sugar solutions for threshold test [68]

Solution No.	Concentration g/100ml	ml of stock solution in water up to 500 ml
1	0.00	
2	0.05	1.25 ml/500ml
3	0.10	2.50
4	0.20	5.00
5	0.30	7.50
6	0.40	10.00
7	0.50	12.50
8	0.60	15.00
9	0.80	20.00
10	1.00	25.00

Stock solution: 50 g sucrose in water up to 250 ml = 20g/100ml

Table 23. Preparation of salt solutions for threshold test [68]

Solution No.	Concentration g/100 ml	ml of stock solution in water up to 500 ml
1	0.00	
2	0.02	1.0ml/ 500 ml
3	0.04	2.0
4	0.06	3.0
5	0.08	4.0
6	0.10	5.0
7	0.12	6.0
8	0.14	7.0
9	0.16	8.0
10	0.18	9.0
11	0.20	10.0

Stock solution: 25 g Sodium chloride in water up to 250 ml = 10/g100ml

Table 24. Preparation of citric acid solutions for threshold test [68]

Solution No.	Concentration g/100 ml	ml of stock solution in water up to 500 ml
1	0.000	
2	0.005	2.5 ml/500 ml
3	0.010	5.0
4	0.013	6.5
5	0.015	7.5
6	0.018	9.0
7	0.020	10.0
8	0.025	12.5
9	0.030	15.0
10	0.035	17.5

Stock solution: 2.5 g citric acid in water up to 250 ml = 1g/100ml

Table 25. Preparation of caffeine solutions for threshold test [68]

Solution No.	Concentration g/100ml	ml of stock solution in water up to 500 ml
1	0.000	
2	0.003	3 ml/500ml
3	0.004	4
4	0.005	5
5	0.006	6
6	0.008	8
7	0.010	10
8	0.015	15
9	0.020	20
10	0.030	30

Stook solution: dissalve 1.25 g coffeine in hot (50–60 °C) water up to 250 ml (20 °C) = 0.5 g/100 ml

5 Microbiological Analysis

5.1 Facilities, Equipment, Glassware, and Media for a Modest Microbiological Laboratory [78–86]

A well-ventilated and illuminated separate room should be used exclusively for microbiological evaluation. An ambient temperature of 21 to 25 C and relative humidity of about 50% are recommended. Direct sunlight is known to have deleterious effects on reagents, media and specimens.

The benches should have a surface of impervious material so that they can be cleaned easily prior to use. The ideal bench top height is 90 to 95 cm with a depth of 75 cm. Media preparation and glassware cleaning areas should be separated from the analytical areas. Floors, walls and ceiling should be covered with material which provides a smooth impervious surface easily disinfected. Good-grade enamel or epoxy paint are recommended. Eating, smoking and unnecessary traffic should never be permitted in the laboratory.

Storage space in the laboratory should be sufficient for a cabinet, a refrigerator and a freezer. The laboratory should be equipped with gas jets for bunsen burners, water supply, electrical outlets, adequate number of sinks with hot and cold tap water and a waste disposal system. The following equipment is essential:

- A steam autoclave operating at 15 psi for the sterilization of liquid media. For checking the proper functioning of the autoclave, *Bacillus stearothermophilus* spore ampules or strips, are available from several commercial sources.
- A hot-air sterilizing oven (160 C) for the sterilization of glassware.
- Agar-tempering water bath thermostatically controlled (45–50 C).
- Two incubators, one set at 35 C and the other at 55 C.
- Microscope with oil immersion lens, glass slides and cover slips.
- Mechanical blender (such as Waring blender).
- Thermometers.
- Two balances, a laboratory balance sensitive to 0.1 g with a 200 g load and an analytical balance having a sensitivity of 1 mg with a 10 g load.
- Transfer needles and loops.
- Test tubes with screw caps.
- Test tube racks.
- Petri dishes – glass or plastic disposable (15 x 100 mm).
- Petri dish boxes (stainless steel or aluminium).
- Pipettes graduated, 1 ml.

- Dilution bottles graduated, 100 ml.
- Wax pencils.
- Beakers, 250 ml, 500 ml, 1000 ml and 2000 ml.
- Pipette boxes (stainless steel or aluminium).
- Erlenmeyer flasks, 250 ml, 500 ml, 1000 ml and 2000 ml.
- Graduated cylinders, 100 ml, 250 ml, 500 ml and 1000 ml.
- Cotton or foam stoppers for plugging the flasks and tubes.
- Colony counter (Quebec or equivalent) or hand tally counter.
- Bunsen burner.
- Refrigerator at 0–5 C.
- Most of the media and reagents used in food microbiology laboratories are purchased prepared or in dehydrated form and should be kept in stock.

Media Preparation

Microbiological media should be prepared with distilled water and sterilized according to instructions given by the manufacturer, usually at 15 psi (121 C) for 20 min. Before use put it in a tempering water bath (45 – 50 C). Dilution bottles should be filled to 102 ml with distilled water, preferably buffered with phosphates, so that a 99 ml volume will result after sterilizing in an autoclave. Dilution bottles should be loosely closed during sterilization, and should be allowed to cool prior to use. Overheating during preparation and sterilization, or resulting from holding too long in the molten state before dispensing into plates or tubes, should be avoided. This can result in loss of productivity through hydrolysis of agar, caramelization of carbohydrates, lowering of pH, increase in inhibitory action, loss of dye content in selective or differential media, and formation of precipitates. Media containing dyes should be protected from light. Glassware and utensils should be free from detergent residue, and should be sterilized in advance at 160 C for one hour and allowed to cool to room temperature prior to use.

Preparation of Phosphate Buffer Solution: A stock solution for preparing dilution water is made by adding 34 g of monobasic potassium phosphate, KH_2PO_4 to 800 ml of distilled water, adjusting the pH to 7.2 with 2 N NaOH and making to 1 l with water. 1.25 ml of this stock solution is added to 1 l of distilled water and dispensed into dilution bottles.

Preparation of Peptone Water Solution: A Solution containing 0.1 % peptone is usually recommended as diluent. 1.5 g of commercial powder containing peptone (1 g) and sodium chloride(0.5 g) is added to 1 l of distilled water. Sterilize by autoclaving at 121 C for 15 min. The pH value is 7.2 ± 0.2.

Preparation of Sodium Chloride Solution (Saline Solution): Dissolve 1 saline tablet in 500 ml of distilled water in order to obtain 0.85% ("normal", physiological, or isotonic) saline. Such a solution has the same osmotic pressure as microorganisms.

Preparation of Buffered Saline Solution: Dissolve 8 g NaCl, 0.34 g potassium dihydrogen phosphate (KH_2PO_4) and 1.21 g dipotassium hydrogen phosphate (K_2HPO_4) in distilled water up to 1 l.

Preparation of Ringer Solution: Ringer solution is ionically balanced solution and it is better than saline solution for making bacterial suspensions. To prepare full strength Ringer solution, dissolve sodium chloride 9 g, potassium chloride 0.42 g, Calcium chloride 0.48 g, and sodium bicarbonate 0.2 g in distilled water up to 1 l. The solution is used at one quarter the original strength, by dilution of 250 ml of this solution with 750 ml of water.

5.2 Microbiological Examination of Tropical Fruit Products

The examination of tropical fruit products for the presence, types and numbers of microorganisms is a normal quality control procedure. In spite of the importance of this, none of methods in common use permit the determination of exact numbers of microorganisms in a fruit product. Although some methods of analysis are better than others, every method has certain limitations associated with its use [78–86].

5.2.1 Direct Microscopic Count (DMC)

Smears of fruit product specimen, or dilutions, are prepared on special slides, stained with an appropriate dye preparation, microbial cells (dead and viable, individual and/or clumps) are counted in a given number of microscopic fields, and the number of organisms/g or /ml are determined by use of the microscope factor. This principle applies to films or counting chambers.

The advantages of the direct microscopic count are that it is rapid, it requires a minimum amount of equipment, the slides may be kept for further identification or reference, and that morphological and Gram-stain types may be evaluated. The disadvantages are that, it is suitable only for fruit products which contain large numbers of microorganisms (high sample error for total cell counts less than $10^5 \, \mathrm{ml}^{-1}$) and that debris and analyst fatigue reduces the precision of the test.

5.2.1.1 Film Method

This method involve the spreading of 0.01 ml of fruit juice or other material over 1 cm^2 of clean slide and allowing it to air dry or drying it over gentle heat. Flood the smear with methylene blue stain (made by dissolving 1 g methylene blue in 100 ml water) for 1 min. North's stain (made by dissolving 3 ml aniline oil in 10 ml 95% ethyl alcohol, and slowly adding while stirring 1.5 ml concentrated HCl and cooling the mixture, then 30 ml methylene blue solution saturated alcohol with methylene blue is added and the solution made up to 100 ml with water) can also be used.

Remove the excess stain by washing the slide in a gentle stream of tap water or immersing the slide in a beaker containing tap water. The slide is then dried and viewed under an oil-immersion microscope. The following technique has been found easier to use:

Mix 5 ml of fruit juice with 5 ml of Hill's stain (make by adding 75 mg crystal violet to 100 ml of distilled water and filter, store in refrigerator) in a test tube. Stopper and shake. Pipet 0.01 ml of mixture to 1 cm^2 of clean slide. Dry prepared slide under a heat lamp. Coat surface liberally with plastic spray such as Krylon. After air drying view under an oil-immersion microscope, magnification is approximately 900 X.

From 10 to 100 fields are usually counted depending on the cell density in the film and the precision desired. To determine the microbial population in a fruit product or ingredient (sugar solution, added water ...) estimate the average number of microorganisms per field, multiply by the Microscopic Factor (MF), and if dilution has been carried out multiply by the reciprocal of the dilution. Report the result as the direct microscopic count (DMC) per g or per ml.

Microscopic Factor (MF)

Microscopes used for this procedure should have microscopic factor between 300 000 and 600 000. Field diameters of 0.2 to 0.15 mm will give these microscopic factors. To determine the microscopic factor, adjust the illuminator to provide maximum optical resolution. Place a stage micrometer ruled in 0.1 and 0.01 mm divisions on the stage of the microscope, and with the 1.8 mm oil immersion objective measure the field diameter in mm to the third decimal place (e.g. 0.146). To determine the area of the field, square the radius (r = 1/2 field diameter) and multiply by, (e.g. (r^2 = 0.005329 x 3.1416 = area of field in square millimeters). To convert the area of one field in mm^2 to cm^2, divide field area in mm^2 by 100. To determine the number of such fields in 1 cm^2, divide by the area (in cm^2) of one field. Since only 0.01 ml of sample is spread over the 1 cm^2 area, multiply the number of fields by 100 to determine the number of fields per ml of dilution used to make the film. The value thus obtained is the microscopic factor and can be calculated as follows:

$$MF = \frac{10\,000}{3.1416 \times r^2}$$

Gram Stain

Gram stain is a differential stain incorporating two dyes of contrasting colors devised in 1847 by Christian Gram, hence its name. Bacteria are classified as either gram positive or gram negative depending upon whether they retain or lose the primary stain (crystal violet) when subjected to a decolourizing agent. The gram stain procedure is as follows:

- Prepare from the fruit juice a smear as previously described.
- Flood with crystal violet solution for 30 seconds. Preparation of Crystal violet solution: Solution A made up of 2 g crystal violet (85% dye content and 20 ml ethyl alcohol (95%).
  Solution B made up of 0.8 g ammonium oxalate and 80 ml distilled water.
  Dilute solution A about 1 in 5 with distilled water and mix with an equal volume of solution B.
- Wash with tap water.
- Flood with iodine solution (made up of 1 g iodine crystals, 2 g of potassium iodide, and 300 ml of distilled water) for 30 seconds.
- Wash with tap water.
- Decolourize by flooding with 95% ethyl alcohol until all but the thickest parts of the smear have ceased to give off dye.
- Counterstain with safranin (made up of 25 ml of 2.5% solution of safranin in 95% alcohol, and 75 ml of distilled water) for 30 seconds.
- Wash with tap water and blot dry.
- Examine with oil immersion objective (98X) and the 10X ocular.

Gram positive bacteria retain the crystal violet stain while the Gram negative bacteria are stained with safranin (rose-colored). In fruit products, Gram positive bacteria are the rule (*Lactobacillus, Bacillus, Streptococcus, etc.*).

5.2.1.2 Counting Chamber

A counting chamber is a glass slide (such as Petroff-Hauser-slide) so constructed that when the chamber is charged with a food suspension and a cover slip placed over it, a definite volume of the suspension can be related to a specific area of the microscopic field. The slide may be ruled to define volumes and the number of organisms per ruled area is counted. The organisms may be stained or counted with dark field illumination. After the counting chamber has been charged and cell motion is minimal, the microorganisms are counted in 25 squares at magnifications of 600 X. The depth of a bright line hemocytometer cell is 0.1 mm, and the rulings cover an area of 1 mm^2. The depth of the counting cell in the Petroff-Hauser-chamber is 0.02 mm, and each square is $2.5 \times 10^{-4} cm^2$ representing a volume of 5×10^{-8} for each square observed. Therefore, a microbial density in excess of 10^7/ml is essential.

Determine the number of organisms per unit volume and multiply by the reciprocal of the dilution used. More information about direct microscopic count and about the Howard mould count and Geotrichum count can be obtained from the latest edition of: AOAC [8] and Speck [86]

5.2.2 Colony Count Methods

Aliquots of fruit product samples are blended, serially diluted in an appropriate diluent, plated in a suitable agar medium, incubated at an approriate temperature for a given time, and all visible colonies are counted. The measure of the total number of aerobic mesophilic organisms initially present in the fruit product, known as the *Standard Plate Count (SPC)* can provide a useful tool for estimating microbial populations in fruit products. The optimum medium and conditions for determining the colony count may vary from one product to another. However, once the best procedure for a given food is determined, it can be very useful for routine microbial analysis.

Tropical fruit products are generally acidic, with pH values less than 4.3. All contain sugars with amounts varying from 8 to 70%. The microorganisms of greatest significance in tropical fruit processing are the lactic acid bacteria (species of Lactobacillus and Leuconostoc). The most frequently found yeasts in tropical fruit products are Saccharomyces, Candida and Torulopsis. Fruit juice concentrates and jams are subject to spoilage by osmophilic yeasts. Therefore, tropical fruit products are examined using the following media:

– *Plate count agar* for total mesophilic bacteria,
– *Orange serum agar* for lactic acid bacteria and yeasts,
– *Dextrose tryptone agar* for thermophilic bacteria,
– *Potato dextrose agar* for yeast and mould,
– *Malt extract agar* for total yeast and mould,
– *Honey medium* for osmophilic yeasts.

5.2.2.1 General Guidelines

– If the count is expected to be in the range of 2.5×10^3 to 2.5×10^5 per ml or g, prepare plates containing 1: 100 and 1 : 1000 dilutions.
– Duplicate plates, per dilution should be employed.
– The serially diluted samples are plated in *(Pour Plate Method)* or onto *(Spread or surface Plate Method)* a suitable agar medium. The pour plate method is the usual method for the Standard Plate Count. One ml of the dilutions is pipetted in the Petri plate and then about 10 ml of the liquified medium (at 45 C) is introduced. As each plate is poured, the medium with the test portions is mixed by rotating the plate first in one direction and then in the opposite direction.

- The use of the *Spread Plate Method* usually results in higher count than are observed with the normal *Pour Plate Method* and has certain advantages. In this method a 0.1 ml sample is spreaded on the surface of the agar medium (the surface should be dried at room temperature for 24 h) with a sterile bent glass rod (hockey stick). The organisms are not exposed to the heat of the melted agar medium. On the other hand, since relatively small amounts of the sample must be used (less than 0.5 ml), the method may lack precision for samples containing few microorganisms.
- The range of 25 to 150 colonies to a plate is the best for counting.
- Microbial cells often occur as clumps or groups in foods. Consequently, each colony that appears on the agar plates can arise from a clump of cells or from a single cell, and should be referred to as a *Colony Forming Unit* (CFU). Report the results as CFU per g or ml.
- Do not prepare or dispense dilutions, or pour plates in direct sunlight.
- If no colonies appears on the 1 : 100 dilution, report the count as "less than 100 (< 100) Estimated Colony Count" as CPU Test.
- Evaluation of colony count in fruit juices usually results in the presence of a considerable number of fruit particles in the plate which makes it difficult to distinguish the colonies easily for accurate counting. This problem can be overcome by adding 1 ml of 0.5% (w/v) triphenyltetrazolium chloride (TTC) per 100 ml of melted agar medium just prior to pouring the plates. Most bacteria form red colonies with TTC. The TTC should be prepared as an aqueous solution, sterilized by passage through a sterilizing filter, and protected from light and excessive heat to avoid decomposition.
- After incubation at 35 C for 48 hours the colonies should be promptly counted. If impossible to count at once, the plates may be stored at 0–5 C for 24 hours.
- For certain fruit juices (clear juices) or for sugar solutions, water and other liquids the Membrane Filter Method can be used. The liquids are passed through a bacteriological membrane filter, pore size 0.4 μm, and plated onto the surface of the chosen agar medium. Examples of the filtration apparatus are given in Sect. 6.3.3.
- See also the guidelines given in Chapter 6 and 7:
  Sect. 6.3 Microbiological examination of water,
  Sect. 7.3. Evaluation of Cleanliness and Sanitation.

5.2.2.2 Culture Media

All media recommended for use in tropical fruit processing plants can be obtained commercially in dehydrated form. If such media are used, they should be prepared in accordance with manufacturer's instructions. Addresses are given in the appendix..

All prepared media, wheter in plates or tubes, should be stored at between 2 to 5 C in moisture-proof containers, such as plastic bags to avoid water loss.

Plate Count Agar (Standard Methods Agar):

Tryptone	5.0 g
Yeast extract	2.5 g
Glucose	1.0 g
Agar	15.0 g
Distilled water	1.0 liter

Dissolve all ingredients in water by boiling. Adjust to pH 7.1. Dispense into tubes or flasks and autoclave at 121 C for 15 min. Final pH should be 7.0 ± 0.1.

To make plate count agar for acid producing bacteria, add 0.04 g bromocresol purple per liter of medium. Incubation is for 48 h at 32 or at 35 C for the standard plate count. Plates may be also be incubated at 45 or 50 C for 48 h or at 5–7 C for 7–10 days.

This medium is a general purpose medium suitable for counting of total microorganisms, for the isolation of mesophilic microorganisms. In the broth (prepared without agar), the growth of microorganisms is indicated by the change of colour from purple to yellow. In plates, the colonies of acid producing bacteria are set off by a yellow halo in a field of purple.

Orange Serum Agar

Tryptone	10.0 g
Yeast extract	3.0 g
Dextrose	4.0 g
Dipotassium phosphate	2.5 g
Agar	17.0 g
Cysteine	0.001 g
Orange serum	200.0 ml
Distilled water	800.0 ml

Prepare orange serum by heating 1 l of freshly extracted orange juice or reconstituted frozen orange juice concentrate to about 93 C. Add 30 g of filter aid and mix thoroughly. Filter under suction using a Buchner funnel. Precoat the filter with the filter aid. Discard the first few milliliters of the filtrate.

Dissolve ingredients in distilled water and sterilize 15 minutes at 121 C. After sterilization, the pH value should be 5.5.

This medium is suitable for cultivating acid-tolerant microorganisms, such as Lactobacillus, Leuconostoc, Bacillus coaguland, B. Thermoacidurans. It is also suitable for yeasts.

Orange serum broth can be prepared as Orange Serum Agar except omitting agar.

Dextrose-Tryptone agar

Tryptone	10.0 g
Dextrose	5.0 g
Agar	12.0 g

| Bromocresol purple | 0.04 g |
| Distilled water | 1.0 l |

Mix the ingredients and boil to dissolve completely. Sterilize by autoclaving at 121 C for 15 min. The pH should be 7.0. Dextrose-Tryptone medium is suitable for isolation of mesophilic and thermophilic organisms in fruit products. It most suitable for the cultivation and enumeration of the thermophilic bacteria causing "flat-sour" spoilage of canned fruit products. It is recommended for the evaluation of canned fruits, dehydrated fruits, and sugar for flat-sour bacteria of Bacillus stearothermophilus type, for mesophilic aerobes and for facultative anaerobes.For the enumeration of mesophilic microorganisms the plates are incubated at 32 C for 72 h. The total number of colonies, with separate totals for acid producing (yellow halo) and non-acid producing colonies can be counted. For the flat-sour bacteria the plates are incubated at 35 C for 48 h. The colonies are typically round, 2–5 mm in diameter, and contain a typical "spot" in the centre (opaque centre), and surrounded by a yellow zone in contrast to the purple medium.

A count of aerobic bacterial spores (Bacillus sp.) can be obtained as follows: Transfer 10 ml of fruit juice with a sterile pipette to a sterile test tube. Heat the tube at 80 C for 15 min using thermostatically controlled water bath, so that the vegetative cells are killed and the surviving spores are heat shocked to promote germination. Cool the tube promptly in water and flame its upper portion to kill any bacteria that may have lodged there. Place 1 ml in each of duplicate plates and pour with Dextrose-Tryptone agar and incubate one at 32 C, for mesophiles, and other at 55 C, for thermophiles. Count the colonies after 48 or 72 h.

Potato-Dextrose agar

Potato extract	4 g
Dextrose	20 g
Agar	15 g
Distilled water	1 l

Bring the mixture to dissolve completely. Sterilize by autoclaving at 121 C for 15 min. The pH should be 5.6 ± 0.2. In order to suppress bacterial growth it is sometimes desirable to acidify the medium to pH 3.5 (or to the pH of the fruit product). This can be done by adding 1 ml of lactic acid 10% to each 100 of sterilized medium at 50 C. Sterile acid solution is added to the flask after the agar has been sterilized and just prior to being poured onto plates. The medium must not be heated after the addition of acid, this would result in hydrolysis of the agar and destroy its gelling properties. Incubate for 5 days at 21 C. Count the number of yeast and mould colonies.

If only a yeast count is desired, 0.25% of sodium propionate is added to the medium, to inhibit mould growth. If a non-selective medium is required, it is suggested that Potato-Dextrose agar may be used without added acid or, alternatively, one may use a general purpose mycological medium such as Malt Extract agar.

Malt extract agar

Malt extract	17 g
Mycological Peptone	5 g
Agar	15 g

Suspend the mixture in 1 liter of distilled water and boil to dissolve and sterilize by autoclaving at 115 C for 10 min. The pH value should be 5.4 ± 0.2.

If it is desired to adjust the pH to 3.5, cool to 55 C and add 2 ml 10% lactic acid solution to every 100 ml of the malt extract agar. Once acidified with lactic acid, the medium should not be re-heated.

Malt extract agar is the recommended medium for the detection, isolation and enumeration of yeast and mould.

Malt extract agar can be also prepared by mixing 30 g malt extract and 15 g agar in 1 l distilled water. Sterilize at 121 C for 15 min and do not over-autoclave this medium. Final pH is 5.5 ± 0.2. The pH can be adjusted to 3.5 by using 10% lactic acid solution.

Malt extract broth can be prepared as Malt Extract Agar, except omitting agar. The medium can be distributed into tubes having Durham tubes for gas production test.

Honey medium

Honey	800.0 g
Peptone	1.0 g
Dipotassium phosphate	1.0 g
Magnesium sulphate	0.5 g
NaCl	0.1 g
Ammonium tartrate	0.5 g
Calcium chloride	0.1 g

Warm the honey, mix the ingredients, and add water. Adjust pH to 4.2 and add water in sufficient quantity to make up to 1 l. Distribute the medium into tubes. Place one Durham tube into each. Plug with cotton wool and sterilize at 121 C for 15 min.

Osmophilic yeast can be indicated by gas production in Durham tube, after incubation at 37 C for 48–72h.

Addresses of commercial suppliers of media and ingredients: see Appendix

5.2.3 Fermentation Test

The Fermentation test is a routine microbiological testing procedure extensively used in fruit processing plants. The fruit product may be examined for sterility or for keeping quality. In this instance the interest is only in ascertaining whether any viable microorganisms are present in the fruit product and to study the fermentative characters of the spoilage microorganism, if it produces gas, acids and/or off-flavor substances. When gas is produced from fermentable substrates by growing microorganisms (mainly yeast and bacteria), it is evidence of fermentation and acid formation. However, fermentation often occurs without gas production, and then acid formation and also possible cloud formation are our evidence that the organism has metabolized the substrate [78,84–86].

Unopened containers (finished products) may be placed at:

35 °C for 15 days, or
55 °C for   7 days.

They are periodically inspected and at the end of the incubation period are examined for evidence of growth:

– formation of gas (swelling),
– formation of lactic acid (see Sect. 5.6.2),
– formation of diacetyl (see Sect. 5.6.2),
– development of off-odors, and/or
– changes in texture, cloudiness and colour.

Such an *Incubation Test* can be also done at room temperature (normal storage) for 21 days.

A preferred method for the *fermentation test* is to transfer about 1 ml of fruit juice or homogenate or 1 g of solid product aseptically into test tubes containing appropriate culture media with a Durham tube, and incubating the tubes at 35 C for 3 days (or at 25 C, for more than a week).

The media used for such a test can be a sugar solution of 9.6 Brix (usual test in the fruit industry) or appropriate culture media (orange serum broth or malt extract broth).

If sugar solution is used as a medium, prepare the solution using sugar commonly used in the production, and dispense 90 ml into 200 ml MEDICAL PLATES with screw tops. An inverted Durham tube (75 x 10 mm) should be placed to each plate. Add paraffin oil to form a 1 cm thick layer on the surface of the sugar solution. Ensure screw tops are loosened and autoclave at 121 C for 15 min. Cool and screw the tops down.

Transfer 10 ml of the fruit juice to the sugar solution below the paraffin layer. Swirl (do not shake) to mix the content, and incubate at 25 C for 10 days or more, examining daily for gas production in the Durham tube and paraffin layer.

As soon as the positive reaction is confirmed cease the incubation and record the result and day of incubation. The presence of yeast and/or bacteria can be checked by microscopic examination. Acid production in the culture medium may be detected by adding to the medium an indicator or dye such as bromo-cresol purple, which changes from purple to yellow in the presence of acid.

Mould growth should be absent at all times due to the layer of paraffin. However, if any mould colonies are noted but the test is otherwise negative, record as negative.

Typical spoilage organisms in tropical fruit products:

− *Yeasts:*
  Strong fermenters (Gas + Cloud)
  Saccharomyces cerevisiae, S. uvarum, S. bailii
  Weak fermenters (Gas + Cloud)
  Candida, Pichia, Hansenula, Kloeckera, Brettanomyces, Hanseniaspora.
  Cloud formation
  without gas: Rhodotorula.
− *Bacteria:*
  Bacilli, formation of gas + cloud:
  Lactob. buchneri, Lactob. confusus.
  Bacilli, formation of cloud without gas:
  Lactob. casei, Lactob. plantarum.
  Cocci, formation of cloud + slime + gas:
  Leuconostoc mesenteroides, Leuconostoc dextranicum.
  Lactobacillus and Leuconostoc are gram positive and producing lactic acid.

5.3 Classification of Microorganisms

Classification of microorganisms, or taxonomy, is the systematic arrangement of microorganisms in named groups or categories according to some definite scheme, property or group of properties. The quality assurance laboratories may need to know only the group of the microorganism (such as lactic acid bacteria) and in some cases name of the genus (such as Lactobacillus).

Certain types of bacteria may be distinguished by the appearance of their colonies in a Petri dish. Acid producing bacteria will cause the Bromo Cresol Purple indicator in the medium to turn yellow in a zone around the colonies, which can be counted and reported separately. On a orange serum agar, the small size "pin-point" subsurface colonies are Lactobacillus sp., Leuconostoc sp. grow as large viscous colonies, stand out from the surface of the medium and are opalescent and gummy in character.

Aerobic sporeformers (Bacillus sp.) form colonies that have a tendency to spread on the surface of the plate. The thermophilic "flat-sour" sporeformers (such as Bacillus coagulans and B. stearothermophilus) give colonies with a dark center and a yellow halo in the Dextrose-Tryptone agar.

Using a system classification (such as Bergey's Manual [87] or Faddin [88]) and traditional methods, microbiologists can identify the species of the bacteria within days.

The study of the yeasts is more diffucult. Lodder-Taxonomy [89] is generally used. Identification to generic level depends on microscopic observations and includes biochemical, physiological and particularly reproductive characteristics that can take several weeks.

5.4 Detection of Microbial Contamination and Spoilage Using Chemical Methods

There is a need for simple, rapid, accurate, cheap and sensitive methods of monitoring the microbial contamination and spoilage of tropical fruit products. Estimation of the microbial status of fruit products by the traditional methods involves skilled personnel, specialist facilities, and results are only available after at least 72 h.

An alternative approach is to assay the fruit product for the chemical products of microbial growth. Such chemical methods yield more rapid results and have the advantage of identifying spoilage even after the microorganisms have died out. Determination of diacetyl, acetoin, lactic acid, alcohol and other substances have been suggested as indicator of microbial spoilage of fruit products.

5.4.1 Determination of Diacetyl

Diacetyl (2,3-butanedione) in tropical fruit juices is produced by acid forming bacteria (such as *Lactobacillus* and *Leuconostoc*) and yeasts (such as *Saccharomyces*). Determination of diacetyl in fruit juices and concentrates provides a rapid index of sanitation during processing. It is a part of the quality assurance program in the citrus processing industry.

The threshold of diacetyl in water is only 5 ppb and in beer 20 ppb; and it perceived negatively in concentration of more than 350 ppb in fruit juices. It causes citrus fruit juices to have a "buttermilk"-odor if the concentration exceeds 1 ppm.

Diacetyl	$CH_3 - CO - CO - CH_3$
Acetoin	$CH_3 - HCOH - CO - CH_3$

Acetoin (Acetylmethyl-carbinol) is also a fermentation product which produces a coloured derivative with the reagent used for the determination of diacetyl, but is essentially odorless.

The determination of diacetyl and acetoin is valuable at early stages of growth but becomes less useful if bacterial growth is advanced.

The method recommended for the determination of diacetyl and acetoin in tropical fruit juices is based on the distillation method described by Hill [91], or on GLC according to Parish et al. [92].

5.4.1.1 Distillation Method [91]

Equipment:

- *Distillation apparatus,* consisting of 500 ml-flask, connecting tube, water type condenser and 25 ml-graduate cylinders.
- *Hot plate,* 1200 watt.
- *Spectrophotometer,* at 530 nm (Spectronic 20, Bausch and Lomb colorimeter.

Reagents:

- *Alpha-Naphthol solution, 5%,* dissolve 5 g in 100 ml 99% isopropyl alcohol.
- *Potassium hydroxide-creatine solution,* dissolve 40 g KOH pellets in water to make 100 ml of solution. In this, dissolve 1 g creatine. The reagent is not stable. New reagent should be prepared every 3 days and it should be kept at 0–4 C.
- *Standard solutions,* the diacetyl should be freshly distilled (bp 87–90 C) and stored under nitrogen in a freezer. Mix 1 ml of diacetyl in 100 ml of water. Make a 1 : 100 dilution to prepare a standard of 10 ppm. This can be stored up to 4 weeks if kept frozen. Prepare working standards of 0.5, 1, 2, 3, 5 and 7 ppm by diluting the 10 ppm-standard each day by adding 5, 10, 20, 30, 50, and 70 ml to 100 ml. Prepare a calibration curve.

Procedure: Distill 300 ml single-strength juice or reconstituted concentrate at a moderate rate (about 5 ml/min). Collect 4 portions of each of 25 ml of the distillate through filter paper to remove turbidity. About 90% of the total diacetyl content is recovered from the first distillate. Acetoin has a higher boiling point (bp 88–148 C) and the third or the fourth distillate contain most of the acetoin. The results of the third and/or fourth fraction subtracted from the first fraction produces a more accurate value for diacetyl. Transfer 10 ml of distillate and 10 ml of water (for a blank) into two test tubes. Add 5 ml of alpha-naphthol solution and 2 ml of KOH-creatine solution to each tube. Mix the contents in each tube, wait for 5 min and determine the absorbance at 530 nm using the reagent blank to null the spectrophotometer.

Calculation:

Acetoin μg/ml $= d_4\ V/d$.
Diacetyl μg/ml $= (d_1 - d_4)$

where:

d   = volume (ml) of distillate = 25
V   = volume (ml) of sample = 300; i.e., V/d = 300/25 = 12
d_1 and d_4 = the quantities (μg) of diacetyl + acetoin found in the first and fourth 25 ml distillates, respectively.

5.4.1.2 GLC Method [92]

Diacetyl and acetoin are used as indicators of inadequate sanitation during citrus juice processing operation. As an undesirable off-flavour in citrus, diacetyl can be detected organoleptically at concentrations greater than 1 ppm. Although flavourless, acetoin is indicative of microbial growth and can be auto-oxidized to diacetyl. Rapid and acurate monitoring of diacetyl concentration in juice feed streams, fresh juices and imported juice concentrates is critical to maintaining final product quality.

Standard solutions: Prepare a standard stock solution of diacetyl (10 ppm) and acetoin (1000 ppm) fresh daily in distilled water or fruit juice. Prepare standard curves using serial dilutions between 0.05 and 0.5 ppm

GLC conditions: use a gas chromatograph equipped with a flame ionization detector (Perkin-Elmer Sigma 3B), a 1.8 m x.0.32 cm stainless steel 0.3% carbowax 80/100 Carbopack C packed column (Supelco, Bellefonte, PA). The optimum GLC conditions for detection of diacetyl are isothermal injector, detector, and column temperatures at 175 C, 250 C, and 110 C, respectively. The carrier gas is nitrogen at a flow rate of 25 ml/min. The conditions for acetoin are the same except that the column temperature is increased to 135 C and the flow rate of nitrogen is reduced to 15 ml/min.

Sample preparation for GLC: Distill samples (300 ml) of either water or juice containing diacetyl or acetoin using a Scott oil apparatus, AOAC (8). Use the first 2.0 ml of distillate as the sample for GLC analysis. Retention times for diacetyl and acetoin are 2.8 and 6.8 min, respectively. The detection limits for diacetyl and acetoin are 0.05 ppm and 100 ppm, respectively. Elevating the column temperature and decreasing the carrier gas flow rate, increase the limit of detection for acetoin to 10 ppm, but diminished the resolution of the diacetyl peak.

5.4.2 Determination of Lactic Acid

Lactic acid bacteria are the bacterial group chiefly involved in the spoilage of tropical fruit products. Species of the genera Leuconostoc and especially Lactobacillus being commonly found as the causative organisms. The main reason for its predominance is their high tolerance to pH values between 3.5 and 4.5 Other factors that favor the growth of lactic acid bacteria are: their ability to grow very well at low redox potential, their high tolerance for CO_2, and their ability to grow at temperatures used for fruit and fruit-juice processing. Lactic acid bacteria sometimes appear as a secondary infection that is introduced during fruit processing.

Rapid automated techniques based on methods such as HPLC, gas chromatography and radiometry for the determination of lactic acid have been developed as an alternative to traditional microbiological methods for the detection of lactic acid bacteria. But all these techniques suffer the disadvantages of requiring specialized equipment, often of high capital cost, and a high degree of skill in sample preparation. If such equipments are available the methods described in Sect. 2.6.2. (Determination of organic acids using GLC and HPLC) can be used.

For the detection of microbial spoilage in processed tropical fruit a rapid chemical spot test can be recommended as an inexpensive alternative [93]. For the quantitative determination of lactic acid in fruit products the enzymatic evaluation of lactic acid is suitable [16,94,95].

5.4.2.1 A Rapid Spot Test for the Detection of Lactic Acid

The test is based on that described by Ackland and Reeder [93], which does not require solvent extraction of lactic acid, can be completed within 5 min and can be used for highly coloured fruit products which are unsuitable for TLC analysis. The spot test was shown to confirm microbial spoilage detected by cultural techniques. It could also identify samples in which spoilage had occurred, but in which spoilage microorganisms were no longer viable. Advantages of the test are the low cost and its simplicity. The test may not be appropriate for products which have a naturally high lactic acid content [93].

Procedure: Prepare the lactic acid-sensitive swab as follows: assemble the cotton wool swab, as shown in Fig. 20 by inserting the end of a swab through a small hole in a bacteriological seal (Astell). Place approximately 50 mg of ceric sulphate in a glass test tube and mix with ten drops of the sample added with a Pasteur pipette. Immerse the swab briefly in 10% aqueous sodium nitroprusside. Add 1 drop of 10% aqueous piperazine to the swab, with a Pasteur pipette. Add then approximately 200 mg anhydrous calcium chloride to the test tube. Insert the swab assembly, as shown in Fig. 20, so that it is 1–2 cm above the reactants, and seal the test tube with the Astell stopper. Mix the reagents in the bottom of the tube thoroughly by gentle agitation.

For maximum sensitivity and reproducibility the reagents should be added to the test tube and swab in the order described above, and the operation should be completed in less than 5 min.

When lactic acid is present in the sample a deep blue coloration developed on the cotton wool swab within 2 min. Care is taken to avoid contaminating the piperazine with the nitroprusside solution as this cause rapid decomposition of the piperazine, also to avoid contact between the swab and the reagents in the bottom of the tube as false positives may result. The nitroprusside solution is best kept refrigerated, and the piperazine solution in the dark at room temperature. With these precautions both reagents remain stable for several months.

The method described involves the oxidation of lactic acid to acetaldehyde by ceric sulphate ($CeSO_4$), the dissolution of anhydrous calcium chloride ($CaCl_2$) providing heat to drive the reaction. The acetaldehyde produced is detected by its colour reaction with soduim nitroprusside, $Na_2(Fe(CN)_5NO) \cdot 2\ H_2O$, and a secondary amine, in this case piperazine.

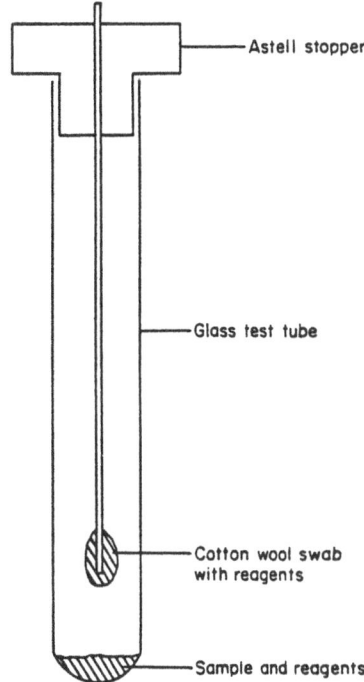

Fig. 20. Equipment for the spot test of lactic acid [93]

5.4.2.2 Enzymatic Determination of Lactic Acid [16, 94, 95]

Principle:

In the presence of L-lactate dehydrogenase (L-LDH), L-lactic acid (lactate) is oxidized by nicotinamide-adenine dinucleotide (NAD) to pyruvate. The oxidation of D-lactic acid requires the presence of the enzyme D-lactate dehydrogenase (D-LDH).

The equilibrium of these reactions lies almost completely on the side of L-lactate. However, by trapping the pyruvate in a subsequent reaction catalyzed by the enzyme glutamate-pyruvate transaminase (GPT) in the presence of L-glutamate, the equilibrium can be displaced in favour of pyruvate and NADH.

The amount of NADH formed in is stoichiometric with the concentration of L-lactic acid and D-lactic acid, respectively. The increase in NADH is determined by means of its absorbance at 334, 340 or 365 nm.

Reagents:

– L-Glutamic acid
– Sodium hydroxide, 10 mol/l
– Glycylglycine
– NAD

- GPT
- L-LDH
- D-LDH
- L-Lactate and D-Lactate standards for the preparation of standard solutions.

Preparation of Solutions (for ca. 50 determinations):

- Buffer (Glycylglycine, 0.6 mol/l; L-glutamate, 0.1 mol/l; pH 10.0): Dissolve 4.75 g glycylglycine and 0.88 g L-glutamic acid in approx. 50 ml water, adjust to pH 10.0 with approx. 4.6 ml NaOH (10 mol/l) and fill up to 60 ml with water. The buffer is stable for at least 3 months at 4 C. Bring buffer solution to 20–25 C before use.
- NAD solution (approx. 47 mmol/l): Dissolve 420 mg NAD in 12 ml water. The solution is stable for at least 4 weeks at 4 C.
- GPT solution (20 mg/ml): Centrifuge 2 ml of the suspension (10 mg/ml) for 10 min at approx. 4000 rpm; suck off 1.0 ml of the clear suspension and discard. The suspension is stable for at least one year at 4 C.
- L-LDH solution (10 mg/ml): Use the glycerol solution undiluted. The solution is stable for at least one year at 4 C.
- D-LDH solution (5 mg/ml): Use the glycerol solution undiluted. The suspension is stable for at least one year at 4 C:
- L-Lactic acid standard solution (2 mmol/l = 0.180 g/l): Dilute L-lactate standard solution, 1 mol/l, 1 to 500 with NaOH 10 mmol/l. Prepare freshly before use.
- D-Lactic acid standard solution (2 mmol/l = 0.180 g/l): Dissolve 49 mg D-lactate, Li-salt, in 250 ml water. Prepare freshly before use.

Procedure: Measurement is made at the absorption maximum of 340 nm when spectrophotometers are used, and at 365 or 334 when a spectral-line photometer with an Hg lamp is used. Use a glass cuvette 1 cm light path at 20–25 C. Read against air (without a cuvette in the light), against water or against blank.

Sample Preparation: Use the method described by the enzymatic determination of citric acid (Sect. 2.6).

More information about the determination of lactic acid in fruit products is given by Boehringer [16].

Table 26. Tests carried out to identify bacteria isolated from mango soft drink preserved by 200 ppm sodium benzoate at pH 3.8 and 3.6 [90]

Property	Isolates	
	pH 3.8	pH 3.6
Cell shape	bacilli	bacilli
Gram reaction	+	+
Spore formation	−	+
Catalase activity	−	+
O_2 requirement	Anaerobic	Aerobic
Nitrate reduction	−	
True branching	−	
MRS Lactobacilli	+	
Colonies Characteristics:		
Form	Round	Round
Internal structure	Granular	Granular
Edge	Entire	Undulate
Elevation	Flat	Flat
Colour	White	White
Consistency	Watery	Watery
Transparency	Opaque	Opaque
Surface	Dull, Smooth	Dull, Smooth
Spread	Non spreading	Spreading
Diameter, mm	0.5 − 1	1.3 − 5
Genera	*Lactobacillus*	*Bacillus*

Table 27. Tests carried out to identify bacteria isolated from mango soft drink preserved by 200 ppm potassium sorbate at pH 3.8 and 3.6 [90]

Property	Isolates	
	pH 3.8	pH 3.6
Cell shape	cocci	cocci
Gram reaction	+	+
Catalase activity	−	−
O_2 requirement	Aerobic	Aerobic
MRS lactobacilli	+	+
Colonies Characteristics:		
Form	Round	Round
Internal structure	Granular	Granular
Edge	Entire	Entire
Consistency	Watery	Watery
Transparency	Opaque	Opaque
Spread	Non spreading	Non spreading
Surface	Smooth	Smooth
Elevation	Flate	Flate
Colour	White	White
Genera	*Streptococcus*	*Streptococcus*

Table 28. Morphological properties of yeasts isolated from mango soft drinks treated with 200 ppm sodium benzoate & potassium sorbate [90.]

Treatment	Morphological properties
Sodium benzoate 200 ppm. pH 3.6 and 3.8	Oval shape multilateral budding, ascospore formation Forms chains, floculates to the top.
Potassium sorbate 200 ppm. pH 3.6	Apiculate (lemon-shaped), from 200 ppm pH 3.6 pseudomycelium, chains of blastospores-arising at the end of pseudomycelium cells.
pH 3.8	Arch shape, budding cells, pseudomycelium, acetic acid odour

Table 29. Physiological tests carried out to identify yeasts isolated from mango soft drinks preserved by 200 ppm potassium sorbate and sodium benzoate. [90]

	Treatment			
	Potassium	sorbate	Sodium	benzoate
Concentration ppm	200	200	200	200
pH value	3.8	3.6	3.8	3.6
Physiological properties:				
1–Fermentative utilization of carbon compounds.				
Glucose	+	+	+	+
Galactose	−	+	+	+
Sucrose	+	+	+	+
Maltose	+	+	+	+
Lactose	−	−	−	−
Raffinose	−	+	+	+
Soluble starch	−	+	+	+
2–Assimilation of carbon compounds.				
Glucose	+	+	+	+
Galactose	−	+	+	+
L-Sorbose	−			
Sucrose	+	+	+	+
Maltose	+	−	+	+
Lactose	−	−	−	−
Raffinose	+	+	+	+
Soluble starch	−	−	−	−
D-Xylose	−	+	−	−
Arabinose	−	+	−	−
D-Ribose	+	+	+	+
Rhamnose	−	+	−	−
Ethanol	+	+	+	+
Glycerol	+	+	+	+
D-Mannitol	−	−	+	+
Lactic acid	+	+	−	−
Succinic acid	+	+	−	−
3-Assimilation of nitrate	+	+	−	−
Species	*Brettanomyces bruxellensis*	*Candida curiosa*	*Saccharomyces cerevisiae*	

6   Water Control

6.1   Importance and Standards

Water is used for several purposes in the tropical fruit processing plants. It is a part of the final product, as in drinks and canned fruits, and universally used for cleaning of equipment and processing lines, for washing of fruits, as a carrier of raw materials between unit processes, as a medium for blanching, preheating. Water is used also for cooling purposes and in boilers and heating systems.

This means, water is in continuous contact with fruits and fruit products during preparation and processing.

The quality of water used throughout the fruit processing plants should at least meet the standard required of drinking water (Table 30). Municipal water systems supply safe drinking water.

For example drinking water in Germany has the following standards [96]:

pH	6.5–9.5	
Iron	< 0.2	mg/l
Nitrate	< 50.0	mg/l
Nitrite	< 0.1	mg/l
Arsenic	< 0.01	mg/l
Cadmium	< 0.005	mg/l
Lead	< 0.04	mg/l
Mercury	< 0.001	mg/l

However, fruit processing plants require more specialized standards than those of some public supplies. Water which is safe to drink is not always suitable for fruit processing. Therefore fruit processing plants often operate their own wells and treatment systems to assure an adequate supply of water with the characteristics desired.

The recommended standard for water used in tropical fruit processing plants is given in Table 30, and will be explained later in Sects. 6.2 and 6.3.

For those who want to know more about water quality, water standards and water treatment several publications are available [97–104].

Table 30. Comparison of recommended quality characteristics of water used by the tropical fruit processing industries and maximum acceptable concentrations of drinking water (WHO International Standard for Drinking Water)

	Recommended concentration[*]	Standard for drinking water
Colour	colourless	< 5 units
Turbidity	clear	< 5 units
Odour	unobjectionable	unobjectionable
Taste	unobjectionable	unobjectionable
Dissolved solids	< 500 mg/l	-----
Alkalinity	< 100 mg/l	-----
Hardness	< 100 mg/l	-----
pH	7.0–8.5	-----
Iron (Fe)	< 0.1 mg/l	< 0.3 mg/l
Nitrate (NO_3)	< 10 mg/l	< 45 mg/l
Arsenic (As)	< 0.05 mg/l	< 0.05 mg/l
Cadmium (Cd)	< 0.005 mg/l	< 0.01 mg/l
Lead (Pb)	< 0.05 mg/l	< 0.05 mg/l
Mercury (Hg)	< 0.001 mg/l	< 0.001 mg/l

[*] Author's recommendations, for explanation see text.

6.2 Methods of Analysis

The methods outlined in this section have been adapted from "Standard Methods for the Examination of Water and Waste Water", American Public Health Association, 16th Edition 1985; except as otherwise indicated and referenced.

Commercial firms have made simplified test kits as well as monitoring systems available for plant operators. This has made the control of water supplies and water treatment easier. Drop test kits make an analysis a simple process. The slide comparator method using the colorimetric method, which compares the sample to be tested against a colour standard, is satisfactory for many plant operations. Water analysis can also be made with a number of instruments currently available.

Some companies producing such kits and instruments are: Fischer Scientific, HACH Company, E. Merck and Taylor Chemicals, Inc. (see Appendix).

Tests generally made for determining water quality in a tropical fruit processing plant are:

– Colour, turbidity, odour and taste.
– Total solids, pH, alkalinity and hardness
– Certain metals and specific conductance.
– Nitrate and residual chlorine.

6.2.1 Measurement of Colour

The ideal for water used in fruit processing industry is that it should be clear and bright in appearance, free from colour (not greater than 5 Hazen or Platinum-cobalt Scale Units), odourless and tasteless and free as possible from organic matter.

Colour in water may result from the presence of natural metallic ions (iron and manganese), plankton, weeds and industrial wastes. Iron in excess of about 0.3 to 0.5 ppm will cause water to appear rusty.

Fruit processing plants will find activated carbon filters useful for improving the colour and the odour of certain water supplies. These filters also absorb phenols and chlorine. Filter media are also available for removing iron and manganese from water.

The term *"colour"* is used herein to mean true colour – that is, the colour of the water from which the turbidity has been removed. The term *"apparent colour"* includes not only the colour due to substance in solution, but also that due to suspended matter. Apparent colour is determined on the original sample without filtration or centrifugation. Colour may be determined by visual comparison of the sample with known concentrations of coloured solution (using platinum-cobalt comparator). Pollution by certain industrial wastes may produce unusual colours that cannot be matched, in this case a spectrophotometric method is recommended.

Platinum-Cobalt Comparison Method: The platinum-cobalt method for measuring colour is given as the standard method, the unit of colour being that produced by 1 mg/l platinum in the form of the chloroplatine ion. The colour value of water is extremely pH-dependent and invariably increases as the pH of water is raised. For this reason, when reporting a colour value, specify the pH at which the colour is determined.

Reagent: Dissolve 1.246 g potassium chloroplatine, K_2PtCl_6 (equivalent to 500 mg metallic platinum) and 1.00 g crystallized cobaltous chloride, $CoCl_2$. 6 H_2O (equivalent to about 250 mg of metallic cobalt) in distilled water with 100 ml conc. HCl and dilute to 1000 ml with distilled water. This stock standard has a colour of 500 units.

Prepare standards having colours of 5, 10 and 15 units by diluting 0.5, 1.0 and 1.5 ml stock colour standard with distilled water to 50 ml in Nessler tubes. Protect these standards against evaporation.

Measurement: Fill the sample in Nessler tubes to the 50-ml mark. Place the sample on a white filter paper sheet and match the colour by looking down vertically. If turbidity is present, report the colour as apparent colour. To determine the true colour in turbid samples, centrifuge the sample till the supernatant liquid is clear and measure the colour as before.

6.2.2 Measurement of Turbidity

Clarity of water used in the fruit processing industry is one of the most important factors to be determined daily. One turbidity unit is considered to be the maximum level permitted. However, up to five units be allowed if it can be demonstrated that the contaminants pose no health hazards or cause no interference with taste or disinfection. It is necessary for a good processing to use water which is clear, colourless and odourless.

Turbidity in water is caused by the presence of suspended matter, such as clay, silt, sand, finely diverse organic and inorganic matter, and biological organisms such as viruses, bacteria, algae, and higher forms of life.

Turbidity is an expression of the optical property that causes light to be scattered and absorbed rather than transmitted in straight lines through the sample. Attempts to correlate turbidity with the weight concentration of suspended matter are impractical because of the size, shape, and refractive index of the particulate materials are important optically but bear little direct relationship to the concentration and specific gravity of the suspended matter.

The suspended particles in water may range in size from 100 000 nm (milli-micron) in diameter for fine sand to collodial suspensions with particle sizes from 1 to 200 nm. Silt with a particle diameter of about 10 000 nm tends to settle out as a sediment. To produce clear water for the fruit processing industry, additional removal of particles in colloidal suspension is usually necessary. Since colloidal suspensions are relatively stable, a coagulant is used to cause aggregation of particles of sufficiently high density to promote settling out for clarification. A coagulant added to water releases positive ions which attract the negatively charged colloidal particles. A jelly-like floc forms which assists in the collection of particles to facilitate clarification. Inorganic chemicals commonly used as coagulants are:

- Ferric sulphate ($Fe_2(SO_4)_3$ Ferrous sulphate $FeSO_4$
- Filter alum $Al_2(SO_4)_3$ Sodium aluminate $Na_2Al_2SO_4$
  (Aluminum sulphate)

The coagulants react with alkalinity in water, resulting in a gelatinous floc in the form of hydroxides. Rapid settling can be achieved by the addition of filter aid. In recent years reverse osmosis have become more popular to facilitate clarification of water to a high degree, especially when used in conjunction with a prefilter and an ion exchanger. The process involved is a simple one as it requires no chemicals, low maintenance requirements, and needs a minimum of energy. A reverse osmosis system not only removes living organisms, such as bacteria and collodial or other nondissolved solids, but also rejects almost all of the dissolved minerals in the water.

Reverse osmosis separates one component of a solution from another by placing the solution under pressure against a semipermeable membrane. Typically the pores of the membrane used in reverse osmosis are 0,5–2 nm in diameter. Molecules which are larger than the pores are rejected by the membrane.

The standard method for the determination of turbidity has been based on the Jackson candle turbidimeter. However, the lowest turbidity value that can be measured directly on this instrument is 25 units. With turbidities of treated water generally falling within the range of 0 to 5 units. Most commercial turbidimeters available for measuring low turbidities give comparatively good indications of the intensity of light scattered in one particular direction, predominantly at right angels to the incident light. The principle of operation in Jackson turbidimeter is based on measuring the length of a sample's light path needed to cause the image of candle flame to disappear, leaving a uniform light field. This is done by adjusting the amount of solution in the calibration glass sample tube. The graduations of the sample tube are then correlated to a turbidity value expressed in JTU (Jackson turbidity units).

Modern nephelometers are relatively unaffected by small changes in design parameters and are therefore specified as the standard instrument for measurement of low turbidities. Other nonstandard turbidimeters, such as the forward-scattering devices, are much more sensitive than nephelometers to the presence of larger particles and are quite useful for process monitoring. Since there is no direct relationship between the Jackson candle turbidity, there is no valid basis for the practice of calibrating a nephelometer in terms of candle units. To distinguish between turbidities derived from the nephelometric and visual methods, the results from the former should be expressed as nephelometric turbidity units (NTU) and from the latter as Jackson turbidity units (JTU).

For the measurement of turbidity follow the manufacturer's operating instructions.

6.2.3 Measurement of pH

The pH of a solution refers to its hydrogen ion activity and is expressed as the logarithm of the reciprocal of the hydrogen ion activity in moles per liter at a given temperature.

The pH of water can generally be related to the mineral composition. Where carbonates and bicarbonates predominate, pH values of water are usually above 7.0.

For control purposes the determination of pH provides reliable information on the potential effect of the water treatment. pH is used in the calculation of carbonate, bicarbonate, and carbon dioxide, corrosion and stability index and other acid-base equilibria of importance to water and wastewater analysis and treatment control.

The pH may be determined by using narrow range indicators or test papers or more accurately by using a pH meter. The glass electrode method is the standard technique.

The colorimetric method is less expensive but suffers from interferences due to colour, turbidity, salinity, colloidal matter, and various oxidants and reductants. The indicators are subject to deterioration. Moreover, no single indicator encompasses the pH range of interest in waters and wastewaters. For these reasons, the colorimetric method is suitable only for rough estimation.

The indicators for the various ranges of pH are given below:

Indicator	pH-range	colour in acid	colour in alkaline
Methyl orange	2.9–4.6	red	orange/yellow
Bromocresol green	3.8–5.4	yellow	blue
Methyl red	4.4–6.0	red	yellow
Bromothymol blue	6.0–7.6	yellow	blue
Cresol red	7.2–8.8	yellow	red
Phenolphthalein	8.3–10.0	colourless	pink
Mixed bromocresol	above 5.2	greenish	blue
green-methyl red	4.6–5.0	5.0 light blue	
	4.8	light pink	grey
	4.6	light pink	

For the measurement of pH using pH-meters see Sect. 2.8.

6.2.4 Measurement of Conductivity

Conductivity is a numerical expression of the ability of a water sample to carry an electric current. It is a collective measure of dissolved ions. This means that the number depends on the total concentration of the ionized substances dissolved in the water and the temperature at which the measurement is made. Most salts (such as sodium carbonate and sodium chloride) are relatively good conductors.

Freshly distilled water has a conductivity of 0.5 to 2 µS/cm, increasing after a few weeks of storage to 2 to 4 µS/cm. This increase is caused mainly by absorption of atmospheric carbon dioxide. The conductivity of potable waters ranges generally from 50 to 1500 µS/cm, and usually less than 400 µS/cm.

Specific conductance values are often used to calculate approximate dissolved solute concentration in mg/l. Figure 21 shows the correlation between specific conductance and gravimetric concentration for six single salts. The curves suggest that the factor by which conductivity should be multiplied to yield total dissolved ion concentration might range from about 0.4 to 0.7 at a conductance of 500 mS/cm. At a conductance of 3000 µS/cm the range would be from 0.5 to 1.0.

Commercially available monitoring equipment are useful for conductivity measurements of water. Conductivity can be recorded either by single-parameter instruments or by more elaborate monitors that also measure and record other variables such as dissolved oxygen, pH and temperature (Fisher Scientific, Cole-Parmer Instrument Company, Kleinfeld Labortechnik, Ciba Corning Analytical (see Appendix)).

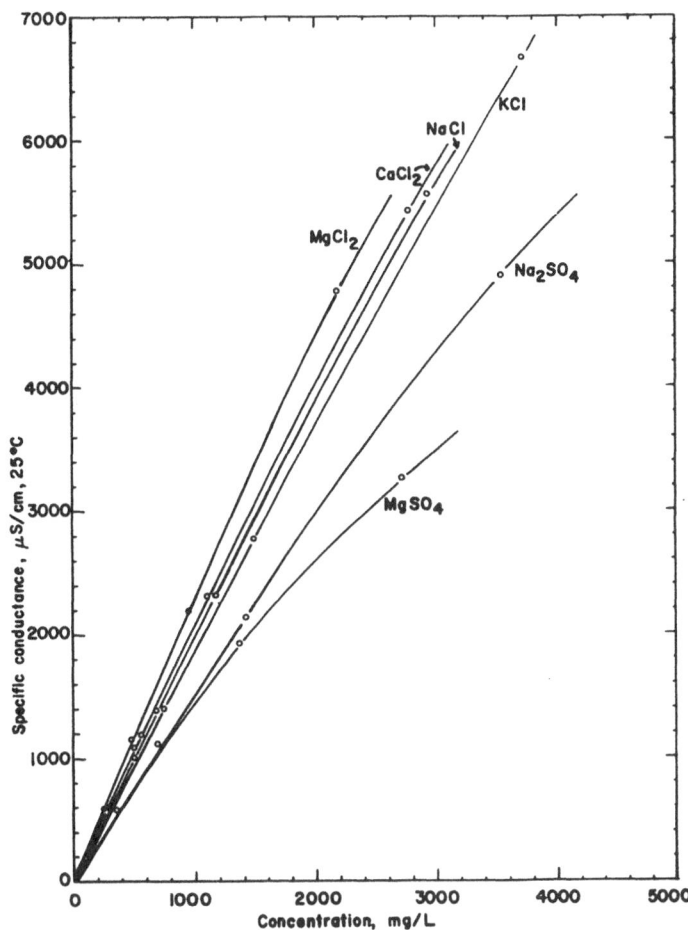

Fig. 21. Specific conductance of solutions of single salts at 25 °C at various concentrations in milligrams per Liter [103]

6.2.5 Determination of Residue, Filterable and Nonfilterable Residue

The term "residue" refers to solid matter suspended or dissolved in water. According to the actual method of analysis, total residue refers to the material left in the vessel after evaporation of a sample and its subsequent drying in an oven at a defined temperature (at 103–105 C or at 180 C). Nonfilterable residue is the portion of the total residue retained by the filter. In the past, this has been referred to as *suspended solids*. Filterable residue is defined as the portion of the total residue that passes through the filter. This has been referred to as *dissolved solids*.

Waters with a high residue are generally of inferior palatability and unsuitable for fruit processing. For these reasons, a limit of 500 mg/l residue is desirable for water used in tropical fruit processing. Waters with very high levels of nonfilterable residues may be aesthetically unsatisfactory. Filterable residues exceeding 500 mg/l will give beverages a brackish taste.

The maximum permissible limits for boiler feed water varies with the operating pressure and is generally 1000 mg/l.

For the determination, evaporate a suitable quantity (about 500 ml) of the well-shaked sample in a tared platinum dish on a water bath. Dry the residue in the oven at 105 C for an hour. Cool in a desiccator and weigh. In expressing the results, state the temperature at which the sample was dried. Residues dried at 105 C may be expected to retain not only water of crystallization but also mechanically occluded water. Residues dried at 180 C will lose almost the mechanically occluded water, but some water of crystallization may remain, especially if sulphates are present. Organic matter is reduced by volatilization but is not completely destroyed. Bicarbonate is converted to carbonate. Some chloride and nitrate salts may be lost. In general, evaporation and drying water at 180 C yield values for total residue that conform more closely to those obtained through summation of individually determined salts than do the values for total residue secured through drying at a lower temperature.

Conductivity measurements are roughly proportional to the filtrable residue.

6.2.6 Determination of Alkalinity

The alkalinity of water is its quantitative capacity to neutralize a strong acid to a designated pH. The German term *"Säurebindungsvermögen"* (acid binding power) express this concept more exactly than the English word "Alkalinity". The measured value may very significantly with the end point pH used in the determination.

Measured alkalinity is usually caused by the carbonates, bicarbonates and hydroxides present in water, the alkalinity is taken as an indication of concentration of these constituents. Carbonate alkalinity is measured by titrating with 0.02 N sulphuric acid using phenolphthalein as the indicator (pH – 8.3). Bicarbonate alkalinity is titrated with 0.02 N sulphuric acid using methyl orange as the indicator (pH – 4.5). Stable pH meters are now commonly available and permit the analysis of portions of titration curve using electronic calculators.

Samples testing over 200 mg/l alkalinity ($CaCO_3$) often require acid rinsing to prevent water spotting and a general dull appearance on equipment surfaces. For tropical fruit processing plants it is often undesirable for the alkalinity to be greater than 100 mg/l.

Reagents

– *CO_2-free distilled water:* Prepare all stock and standard solutions and dilution water for the standardization using distilled water freshly boiled for 15 min and cooled to room temperature.

– *Sodium carbonate solution, 0.05 N:* Dry 3 to 5 g primary standard Na_2CO_3 at 250 C for 4 h and cool in a desiccator. Weigh 2.5 g (to the nearest mg), transfer to a 1 l volumetric flask, and fill to the mark with distilled water.

– *Standard sulphuric acid or hydrochloric acid, 0.1 N:* Dilute 3.0 ml conc H_2SO_4 or 8.3 ml conc HCl to 1 l with distilled water. Standardize against 40.00 ml of 0.05 N Na_2CO_3 solution, with about 60 ml water, in a beaker by titrating potentiometrically to pH of about 5. Lift out electrodes, rinse into the same beaker, and boil gently for 5 min under a watch glass cover. Cool to room temperature, rinse the cover glass into the beaker, and finish the titration to the pH inflection point. Calculate the normality according to

$$\text{Normality, N} = \frac{A \times B}{53.0 \times C}$$

where

A = g Na_2CO_3 weighed into 1 l
B = ml Na_2CO_3 solution taken for titration, and
C = ml acid used.

Use measured normality in calculations or adjust to exactly 0.10 N. A 0.10 N solution = 5.0 mg $CaCO_3$/ml.

– *Standard sulphuric acid or hydrochloric acid, 0.02 N:* Dilute 200 ml of 0.10 N standard acid to 1 l with distilled water. A 0.02 N solution = 1.00 mg $CaCO_3$

– *Phenolphthalein indicator:* Dissolve 1 g in 100 ml ethanol.

– *Methyl orange indicator:* Dissolve 0.5 g in 1 l of water.

– *Mixed bromocresol green-methyl red indicator:* Dissolve 20 mg methyl red sodium salt and 100 mg bromocresol green sodium salt in 100 ml distilled water (or dissolve 20 mg methyl red and 100 mg bromocresol green in 100 ml 95% ethyl alcohol).

– Sodium Thiosulphate, 0.1 N: Dissolve 24.82 g of $Na_2S_2O_3$ x 5 H_2O in 1 l of water.

Procedure: Take an aliquot of the sample requiring less than 50 ml of titrant to get the sharpest colour change at the end point. Remove the free residual chlorine that may be present by adding 0.05 ml (1 drop) of 0.1 N sodium thiosulphate solution or by exposing to ultraviolet light.

– *Phenolphthalein alkalinity:* Pipette 50 ml of the sample into a conical flask. Add 0.1 ml (2 drops) of phenolphthalein indicator. Titrate over a white surface with 0.02 N standard H_2SO_4 until the pink coloration just disappears.

$$\text{Alkalinity, mg/l } CaCO_3 = \frac{A \times N\ 50000}{B}$$

where

A = ml standard acid used, and
N = normality of standard acid.
B = ml of sample

– *Total alkalinity:* This may be determined using either mixed bromocresol green-methyl red indicator or methyl orange alone.
– *Mixed bromocresol green-methyl red indicator method:* Add 0.15 ml (3 drops) of indicator to the solution for which the phenolphthalein alkalinity has been determined or to a fresh sample in a conical flask. Titrate over a wide surface with 0.02 N standard acid to the proper equivalence point is based on the alkalinity concentration as $CaCO_3$ – pH 5.1 for total alkalinity of about 30 mg per litre, pH 4.8 for 150 mg per litre and pH 4.5 for 500 mg per litre. The color response of the mixed indicator is as follows:

above
pH 5.2 greenish blue,
pH 5.0 light blue with lavender gray,
pH 4.8 light pink gray with bluish cast,
and
pH 4.6 light pink

Check the color changes against the reading of a pH meter under the condition of the titration. Report the pH of the end point used as follows: "The alkalinity to pH .... = ... mg/l $CaCO_3$".
– *Methyl orange indicator method:* Add 0.1 ml (2 drops) of indicator to the solution for which the phenolphthalein alkalinity has been determined. Titrate with 0.02 N standard acid until the colour changes from yellow to faint orange at pH 4.6 and pink at 4.0.

$$\text{Total alkalinity} = \frac{(A+B) \times N \times 50000}{C} \text{ mg/l } CaCO_3$$

where

A = ml of acid for phenolphthalein alkalinity,
B = ml of acid for mixed indicator or methyl orange indicator, and
N = normality of standard acid.
C = ml of sample

Calculation of alkalinity relationship: The results obtained from the phenolphthalein and total alkalinity determinations offer a means for the stoichiometric classification of the three principal forms of alkalinity present in water supplies. The classification ascribes the entire alkalinity to bicarbonate, cabonate, and hydroxide.

– *Carbonate alkalinity* is present when the phenolphthalein alkalinity is not zero but is less than the total alkalinity.
– *Hydroxide alkalinity* is present if the phenolphthalein alkalinity is more than the total alkalinity.
– *Bicarbonate alkalinity* is present if the phenolphthalein alkalinity is less than the total alkalinity.
The alkalinity relationship are given in Table 31.

Table 31. Alkalinity relationships[*] [97]

Result of Titration	Hydroxide Alkalinity as $CaCO_3$	Carbonate Alkalinity as $CaCO_3$	Bicarbonate Alkalinity as $CaCO_3$
P = 0	0	0	T
P < 1/2T	0	2P	T–2P
P = 1/2T	0	2P	0
P > 1/2T	2P–T	2(T–P)	0
P = T	T	0	0

[*] Key: P–phenolphthalein alkalinity; T–total alkalinity.

6.2.7 Determination of Hardness

Water hardness is the most common undesirable condition. Most water supplies are hard enough to require softening. Originally, the hardness of a water was understood to be a measure of the capacity of the water for precipitating soap. Soap is precipitated chiefly by the calcium and magnesium ions commonly present in water. Therefore, hardness is defined as a characteristic of water that represents the total concentration of the calcium and magnesium ions expressed as calcium carbonate.

The hardness may range from zero to hundreds of milligrams per liter. The quantitative classification of hardness is as follows:

Condition	mg/l $CaCO_3$
Soft	0–50
Moderately hard	50–100
Hard	100–200
Very hard	over 200

Two types of hardness are encountered temporary hardness and permanent hardness:

– Temporary hardness (carbonate hardness) is due to the presence of calcium and magnesium bicarbonates, which decompose upon heating into insoluble carbonates, thus forming deposits of scale. This is the chief cause of scale formation in boilers which reduce heat transfer efficiency and is a serious problem.
– Permanent hardness (noncarbonate hardness) is due to the presence of soluble calcium and magnesium sulphates, chlorides and nitrates, which are not decomposed by heat and remains after water is boiled, but are precipitated with certain dilute alkalies. Permanent hardness may cause a hard, dense scale on pipelines or equipment which is much more difficult to remove than carbonate scales. An example for such scales is the precipitations in the alkali removal section of a bottle washer, which causes rinsing problems and excessive chain lubrication costs.

Scales on boilers and heat transfer surfaces are due to deposition of insoluble carbonates and sulphates of calcium and magnesium with or without silica. Scaling results in loss in efficiency, reduction in capacity and damage to the equipments.

Scales may harbour bacterial growth, cause excessive soap consumption, the formation of undesirable films such as scums and curds, the plugging of valves, heater tank burnout, failure of thermostate relays, and clog the water system. Corrosion of lines and equipment may also occur. This is especially true of stream condensate lines because of carbonic acid formed by the liberation of carbon dioxide from the bicarbonates.

Most water supplies, especially ground waters, are hard enough to require softening for most purposes. It may be stated that the use of softened water for fruit processing is advantageous. Hardness less than 100 mg/l as calcium carbonate is recommended.

As far as softening of the water is concerned, the current trend appears to be away from significant reduction of water hardness. Around 75–100 mg/l is appropriate. The reason for this trend has been found only recently. Apparently, epidemiological studies have shown a significant increase in cardiovascular disease in the population in areas having very soft water (less than 50 mg/l).

Two methods are employed for water softening treatment. For waters containing 100 to 150 mg/l hardness the lime-soda softening process provides the most suitable and economical treatment. The second method is called an ion-exchange process. This procedure will treat water of low hardness from 50 to 100 ppm. The ion-exchange method is wide application for carbonated soft drink industry. If the initial step in water treatment is solids removal and a reduction in turbidity, some softening is usually achieved as well. For example, sodium aluminate no only acts as a coagulant but adds alkalinity and reduces noncarbonate hardness. The amount of lime required for softening can usually be decreased by 10% when sodium aluminate is used as a coagulant.

In the cold lime softening process, calcium oxide (CaO) is added to the hard water to form calcium hydroxide, which reacts with magnesium and calcium bicarbonates and free CO_2 to form insoluble calcium carbonate and magnesium hydroxide. Magnesium hydroxide is a good flocculating agent which aids in precipitating the calcium carbonate particles. Excess lime is converted to calcium carbonate by the addition of carbon dioxide as the water leaves the primary settling basins. The carbonation at the same time forms magnesium carbonate from magnesium hydroxide. After carbonation, soda ash is added to produce reactions with noncarbonate calcium hardness, converting it to soluble sulphate or chloride.

This treatment will usualy result in water with about 70 to 85 mg/l when discharged from the final filtration unit. Sand and gravel filters are commonly used for removing the precipitated salts. The lime-soda operation is not effective with raw waters of low hardness.

For the determination of hardness, the EDTA titration method, which measures the calcium and magnesium ions is recommended. When the hardness is numerically greater than the sum of the carbonate alkalinity and the bicarbonate alkalinity, the amount of hardness that is equivalent to the alkalinity is called "carbonate hardness"; the amount of hardness in excess of this is called "noncarbonate hardness".

When the hardness is numerically equal to or less than the sum of carbonate and bicarbonate alkalinity, all of the hardness is carbonate hardness, and there is no noncarbonate hardness.

EDTA Titrimetric Method

Ethylenediamine tetraacetic acid (EDTA) and its sodium salts form a chelated soluble complex when added to a solution of certain metal cations. If a small amount of a dye such as Eriochrome Black T is added to an aqueous solution containing calcium and magnesium ions at a pH of 10, the solution will become wine red. If EDTA is then added as a titrant, the calcium and magnesium will be complexed. After sufficient EDTA has been added to complex all the magnesium and calcium, the solution will turn from wine red to blue. This is the end point of the titration. A limit of 5 min is set for the duration of the titration to minimize the tendency toward $CaCO_3$ precipitation.

Reagents

– *Buffer solution*, dissolve 16.9 g ammonium chloride, NH_4Cl, in 143 ml conc ammonium hydroxide, NH_4OH and dilute to 250 ml with distilled water.
– *Indicator*, mix together 0.5 g of Eriochrome Black T and about 100 g of NaCl to prepare a dry powder mixture.
– *Standard calcium solution*, dry $CaCO_3$ at 105 C overnight. Weigh 1 g into a 500-ml conical flask. Add dilute HCl $(1 + 1)$ drop by drop until all the $CaCO_3$ has dissolved. Add 200 ml of distilled water and boil for a few minutes to expel CO_2. Cool, add a few drops of methyl red indicator and adjust to the intermediate orange colour by adding 3 N NH_4OH. Transfer to a 1-l volumetric flask and make up to the mark with distilled water (1 ml = 1 mg $CaCO_3$).
– *Standard EDTA solution*, 0.01 M, dissolve 4 g of disodium salt of EDTA in 800 ml of distilled water. Standardize against standard calcium solution. Adjust the titrant so that 1 ml = 1 mg $CaCO_3$. As the titrant will extract hardness producing cations from soft-glass containers, store in polyethylene (preferable) or pryrex bottles.

Procedure

Select a sample volume that requires less than 15 ml EDTA titrant. Do not extend duration of titration beyond 5 min, measure from the time of the buffer addition. Dilute 25 ml of the water sample with 25 ml of distilled water in a porcelain dish. Add 1–2 ml of buffer solution to give a pH of 10.0 to 10.1. Add an appropriate amount of the indicator. If hardness is present, a red color is formed. Titrate slowly with the standard EDTA solution, stirring continuously, until the last reddish tinge disappears from the solution and a permanent blue colour is produced. In the case of low hardness, take 100–1000 ml sample for titration.

Calculation

$$\text{Hardness (EDTA) as mg/l CaCO}_3 = \frac{A \times B \times 1000}{C}$$

where

A = ml titration for sample
B = mg $CaCO_3$ equivalent to 1 ml EDTA titrant.
C = ml of sample

The hardness is also expressed in terms of grains or degree. The relationship is given blow:

One grain per gallon = 17.1 mg/l (or ppm)
One grain per gallon = One degree of hardness

Simplified Test Kits for Hardness

Commercial firms have made simplified test kits for total hardness determination. An example is the *Merckoquant* Total hardness test strip, and *Aquamerck* Total Hardness Test with precision dropper and indicator solution. The reagents and testing set are manufactured by E. Merck. This method is based on a complexometric titration in which the calcium and/or magnesium ions present in the water form a complex with TITRIPLEX III (ethylenedinitrilotetraacetic acid disodium salt). The indicator also complexes calcium and magnesium ions and is then red in colour. The indicator is liberated on addition of TITRIPLEX III solution whereby its colour changes from red via grey-green to green.

Table 32. Conversion table of the international units used to assess water hardness [105]

	Alkaline earth ions mol/m^3	German degree dH	English degree e	French degree f	ppm CaCO$_3$
1 mol/m^3 alkaline earth ions =	1.00	5.60	7.02	10.00	100.0
1 German degree =	0.18	1.00	1.25	1.78	17.8
1 English degree =	0.14	0.80	1.00	1.43	14.3
1 French degree =	0.10	0.56	0.70	1.00	10.0
1 ppm CaCO$_3$ =	0.01	0.06	0.07	0.10	1.0

6.2.8 Determination of Iron

Most of oxygenated and filtered suface water supplies contain less than 1 mg/l iron. Some ground waters may contain 5 up to 50 mg/l iron. Freshly-drawn ground water containing iron will usually be clear and colourless because iron is in the ferrous form (under reducing conditions). When exposed to air or available chlorine, oxidation converts the ferrous salts (mainly ferrous bicarbonate) into the reddish ferric compounds, which are not significantly soluble. On exposure to air or addition of oxidants, ferrous iron is oxidized to the ferric state and may hydrolyze to form insoluble hydrated ferric oxide. This is the predominant form of iron in water samples. Both forms (ferrous and ferric) are objectionable. Iron in water can cause a bittersweet astringent taste detectable by some people at levels above 1 or 2 mg/l. As little as 0.3 mg/l iron can be expected to cause staining on surfaces of equipment, walls and floors. Iron precipitates with addition of some alkalies or oxidizers such as chlorine. Additional complex phosphates may be required to prevent this problem.

Determination of iron using the atomic absorption spectrophotometric method is accurate, but the phenanthroline method has attained the greatest acceptance for simplicity and reliability. The value of the determination depends greatly on the care taken to obtain a representative sample. Iron in well water or tap samples may vary in concentration and form with duration and degree of flushing before and during sampling. Shake the sample bottle often and vigorously to obtain a uniform suspension of the precipitated iron. For a precise determination of total iron, treat with acid at the time of collection to place the iron in solution.

Phenanthroline Method

Iron is brought into solution, reduced to the ferrous state by boiling with acid and hydroxylamine, and treated with 1,10-phenanthroline at pH 3.2. Three molecules of phenanthroline chelate each atom of ferrous iron to form an orange-red complex. Total, dissolved or ferrous iron concentration between 0.02 and 4.0 mg/l can be determined.

Apparatus:

– *Spectrophotometer* for use at 510 nm, providing a light path of 1 cm or longer.

Reagents: All reagents must be low in iron and use iron-free distilled water.

– *Hydrochloric acid,* HCl, conc., containing less than 0.00005% iron.
– *Hydroxylamine solution,* dissolve 10 g $NH_2OH:HCl$ in 100 ml distilled water.
– *Ammonium acetate buffer solution,* dissolve 250 g $NH_4C_2H_3O_2$ in 150 ml distilled water. Add 700 ml conc (glacial) acetic acid.

- *Sodium acetate solution*, dissolve 200 g $NaC_2H_3O_2$. $3H_2O$ in 800 ml distilled water.
- *Phenanthroline solution*, dissolve 100 mg 1,10-phenanthroline monohydrate, $C_{12}H_8N_2$ H_2O, in 100 ml distilled water by stirring. Add 2 drops conc HCl to the distilled water to facilitate dissolving. One milliliter of this solution is sufficient for no more than 100 µg Fe.
- *Stock iron solution*, add slowly 20 ml conc H_2SO_4 to 50 ml distilled water and dissolve 1.434 g ammonium iron sulphate Fe $(NH_4)_2(SO_4)_2$. 6 H_2O. Add 0.1 N KMn_4 dropwise until a faint pink colour persists. Dilute to 1 l with iron-free water and mix. Each 1 ml will contain 200 µg Fe.
- *Standard iron solution*, prepare daily for use: pipette 50 ml and 5 ml stock solution into two 1-l volumetric flasks and dilute with water. 1 ml = 10 µg Fe.

Procedure:

- *Preparation of calibration curves (0 to 100 µg Fe/100 ml final solution)*, pipette 2, 4, 6, 8, and 10 ml standard iron solution into 100-ml volumetric flasks. Add 1 ml NH_2OH, HCl solution and 1 ml sodium acetate solution to each flask. Dilute each to about 75 ml with distilled water, add 10 ml phenanthroline solution, dilute to volume, mix thoroughly, and let stand for 10 min. Measure the absorbance of each solution in a 5-cm cell at 508 nm against a blank prepared by treating distilled water with the specified amounts of all reagents except the standard iron solution. Correct the absorbance values of standard concentration of iron by substracting the absorbance of blank. From the data obtained, construct a calibration curve for absorbance against mg of Fe.
- *Determination of total iron*, pipette 50 ml of sample into 125-ml erlenmeyer flask. Add 2 ml conc HCl and 1 ml hydroxylamine solution. Add a few glass beads and heat to boiling. Cool to room temperature and transfer to a 50-ml volumetric flask. Add 10 ml ammonium acetate buffer solution and 2 ml phenanthroline solution, and dilute to mark with distilled water. Mix thoroughly and allow 15 min for maximum colour development. Measure the absorbance at 510 nm using a 5-cm absorption cell for amounts of iron less than 100 µg or 1-cm cell for quantities from 100 to 500 µg. Correct the absorbance of the sample by subtracting the absorbance of a sample blank.

Calculation:

$$mg\ Fe/l = \frac{µg\ Fe}{ml\ sample}$$

6.2.9 Determination of Nitrate

Nitrate generally occurs in trace quantities (less than 5 mg/l NO_3) in surface water supplies, but may attain high levels (up to 100 mg/l) in some ground waters. Nitrate is a serious contaminant when present in amounts of 50 mg/l or more, because it causes an oxygen starvation condition in infants. This condition is known as methemoglobinemias (blue babies).

The max. allowable concentration for drinking water is about 40–50 mg/l NO_3, and the recommended concentration is less than 25 mg and for children only 10 mg.

Although nitrates as such are fairly harmless, any consideration of their toxicity should bear in mind that of nitrites. Nitrates may be converted into the more toxic nitrites by the activity of microorganisms. This reduction is largely uncontrolled. It may occur during food processing and storage or in the human digestive tract. In adults the nitrate is converted to nitrite in the intestine, whereas in infants conversion already occurs in the stomach or duodenum, where nitrite is absorbed especially rapidly. Consequently nitrates are particularly toxic to infants. Therefore, for the dairy industry only 0.5 mg/l NO_3 is allowable.

Fruit juices normally have low levels of nitrates. Using ground water for preparing fruit juices from concentrates may increase such levels. A high nitrate content promotes tin uptake in canned fruit juices and drinks.

For the determination of nitrate the colorimetric method using brucine is recommended. The reaction between nitrate and brucine produces a yellow colour that can be measured at 410 nm. The method is recommended only for the concentration range of 0.5 to 5.0 mg/l NO_3, because above this range anomalous results occur while below this range the sensitivity of the method is poor.

Reagents:

- *Standard nitrate solution,* dissolve 0.7216 g anhydrous potassium nitrate, KNO_3, in distilled water and make to 1 l (1 ml = 0.1 mg of N). To prepare working standard solution, dilute 100 ml to 1000 ml with distilled water (1 ml = 0.01 mg of N).
- *Sodium arsenite solution,* dissolve 5 g $NaAsO_2$ and dilute to 1 l with distilled water. The solution is toxic, avoid ingestion.
- *Brucine-sulphanilic acid solution,* dissolve 1 g of brucine sulphate and 0.1 g sulphanilic acid in about 70 ml hot distilled water. Add 3 ml conc HCl, cool and make up to 100 ml. This solution is stable for several months. The pink colour that develops slowly does not affect its usefulness. Brucine is toxic, take care to avoid ingestion.
- *Sulphuric acid solution,* carefully add 50 ml conc H_2SO_4 to 125 ml distilled water. Cool and keep tightly stoppered to prevent absorption of atmospheric moisture.
- *Sodium chloride solution,* dissolve 300 g NaCl and dilute to 1 l with distilled water.

Procedure:

- *Standard curve,* prepare nitrate standards in the range 0–10 mg/l by diluting the standard nitrate solution with distilled water. Pipette 2 ml of each solution, develop and measure the colour as described under "colour development" given below. Draw the standard curve by plotting absorbance against concentration.

– *Pretratment of sample,* if the sample contains chlorine, remove it by adding 0.1 ml of arsenite solution for each 0.1 mg Cl_2 present, and mix. Add one drop in excess to a 50-ml portion.

– *Colour development,* pipette 2 ml of sample containing not more than 10 mg/l nitrate into a 50-ml beaker. Add 1 ml of brucine-solution using a safety pipette. Add 2 ml NaCl solution. Mix thoroughly by hand and add carefully 10 ml H_2SO_4 solution. Mix thoroughly and then place in a well-stirred bath of boiling water that maintains a temperature of 95 C. After 20 min remove the sample and cool. Measure the colour of the standards and the sample at 410 nm.

– Prepare the standard curve from the absorbance values of the standards (minus the blank) run together with the samples. Correct the absorbance readings of the sample by substracting their "sample blank" values from their final absorbance values. Read the concentration of nitrate directly from the standard curve.

Calculation:

$$\text{mg/litre nitrate nitrogen} = \frac{A\ 1000}{B}$$

where:

A	= mg of nitrate nitrogen
B	= ml of sample
mg/litre NO_3	= mg/litre nitrate nitrogen x 4.43.

Simplified Enzymatic Test for Nitrate Determination

The German company E. Merck developed a semienzymatic assay for the determination of nitrate in water (Bioquant Nitrate, E. Merck. Nitrate is partially reduced to nitrite by nitrate reductase. Nitrite is then measured colorimetrically at 525 nm, after diazotation coupling with sulphanilic acid and N-Naphthyl-ethylene diammonium dichloride. The measuring ranges are between 0.5–10 mg/l nitrate and 5–100 mg/l nitrate. Due to its rapidity, the method permits a high number of samples to be tested. Owing to high sensitivity, the colour reaction used permits the exact measurement of low nitrate concentration.

6.2.10 Determination of Residual Chlorine

Chlorination is the one universally accepted method for disinfection of water. Chlorine is effective in killing disease-causing bacteria, slime and algae and aids in precipitating minerals. Food-processing plants have increasingly been chlorinating water for plant use to improve sanitation. The deposition of organic materials from products being handled, and the growth of microorganisms on them is largely eliminated by continuous chlorination. Odours developing from organic matter on belts, washers, flumes, other equipment, an on the floors are prevented by adequate chlorination. A free chlorine residual of at least 0.2 mg/l must be maintained throughout a water distribution system. In general, the maximum content of active chlorine in drinking water must not exceed 0.3 mg/l.

However, water containing free chlorine is not recommended in the preparation of fruit juices, nectars and drinks. Chlorine odour and flavour may be disturbing in concentrations exceeding 5 mg/l. Chlorine present in water used for preparation of fruit juices from fruit juice concentrates can react with phenolic compounds of the juice producing a "chlorophenolic flavour".

When chlorine has been used in all other operations, but eliminated from juice make-up water used for such products, no objectionable off-flavors have resulted.

Chlorine in water may be present as free available chlorine (in the form of hydrochlorus acid or hypochlorite ion, or both) or as combined available chlorine (Chloramines and other chloro derivatives). Both free and combined chlorine may be present simultaneously.

In a tropical fruit processing plant, chlorine is required to be determined in the water used for sanitation, cooling water, and water used in preparation of products.

The iodometric method is generally recommended for the determination of residual chlorine. In this method, chlorine will liberate free iodine from potassium iodide solutions at pH 8 or less. The liberated iodine is titrated with a standard solution of sodium thiosulphate, with starch as the indicator. The reaction is preferably carried out at pH 3 to 4. The minimum detectable concentration is about 40 µg/l Cl if 0.01 N sodium thiosulphate is used with a 500-ml sample.

Reagents:

- *Acetic acid*, (glacial).
- *Potassium iodide*, KI, crystals.
- *Standard sodium thiosulphate*, 0.1 N, dissolve 25 g $Na_2S_2O_3$ x 5 H_2O in 1 l freshly boiled distilled water and standardize the solution against potassium dichromate after at least 2 weeks storage. Add a few ml of chloroform, $CHCl_3$, to minimize bacterial decomposition and to prolong the storage life. Standardize the solution against 0.1 N potassium dichromate (dissolve 4.904 g anhydrous $K_2Cr_2O_7$ up to 1 l).
- *Standard sodium thiosulphate titrant*, 0.01 N or 0.025 N, stability of this working standard is improved if it is prepared by diluting an aged 0.1 N solution. Add a few ml of chloroform. Standard sodium thiosulphate titrants, 0.01 N and 0.025 N, are equivalent, respectively to 354.5 µg and 886.3 µg available Cl/1 ml.
- *Starch indicator,* to 5 g soluble starch add a little cold water and grind to a thin paste. Pour into 1 l of boiling distilled water, stir and allow to settle overnight.

Procedure:

- *Sample preparation,* chlorine in aqueous solution is not stable and will rapidly decrease, particularly in weak solutions. The determination must be made immediately after sampling, avoiding excessive light and agitation. Select a sample volume that will require no more than 20 ml 0.01 N $Na_2S_2O_3$ (for residual chlorine of 1 mg/l or less, take a 1 l sample).
- *Titration,* place 5 ml acetic acid, or enough to reduce the pH between 3.0 and 4.0, in the flask. Add about 1 g KI and mix with a stirring rod. Titrate away from direct sunlight. Add 0.025 N or 0.01 N thiosulphate from a burette until the yellow colour of the liberated iodine is almost discharged. Add 1 ml starch solution and titrate until the blue colour is discharged. If the titration is made with 0.025 N thiosulphate instead of 0.01, then, with a 1-l sample, 1 drop is equivalent to about 50 µ/l; with a 500-ml sample it is about 100 µ/l.
- *Blank titration,* correct the result of the sample titration by determining the blank contributed by such reagent impurities as (1) the free iodine or iodate in potassium iodide that liberates extra iodine; or (2) the traces of reducing agents that might reduce some of the iodine liberated. Take a volume of distilled water corresponding to the sample used for titration, add 5 ml acetic acid, 1 g KI and 1 ml starch solution.
- *Blank titration A,* if a blue colouration develops, titrate with 0.01 N or 0.025 N sodium thiosulphate until it disappears and record the result.
- *Blank titration B,* if no blue colouration occurs, titrate with 0.0287 N iodine solution until a blue colour appears. Backtitrate with 0.01 N or 0.025 N sodium thiosulphate and record the difference as Titration B.
  Substract Blank Titration A from the sample titration, or, if necessary, add the net equivalent value of Blank Titration B.

Calculation

$$mg/l \; Cl = \frac{A \, N \, 35450}{B}$$

where:

A = Titre
N = Normality of $Na_2S_2O_3$
B = ml of sample

Simplified test kits for chlorine

Commercial firms have made simplified test kits for chlorine determination. Such kits make an analysis a simple process. A good example is the *Chlorotex Test.* The chlorotex reagent and testing set are manufactured by the British Drug Houses Ltd.. Add 50 ml of the water sample to 5 ml of the reagent, mix and allow to stand for 1 min. Compare the colour produced with the tints given on a colour chart supplied with the set.

In the presence of 0.1 to 0.2 mg/l Cl, a pink colour develops which changes to red at 0.5 mg/l, purple at 0.6 mg/l and blue at 1 mg/l; green and brown colours occur when the chlorine content exceeds 1 mg/l. A simplified test kit for chlorine determination in water is also available from the German company E. Merck [105].

6.2.11 Determination of Toxic Metals

The determination of toxic metals in the water supply as a routine check is not necessary. A survey twice a year is sufficient.

The presence of toxic metals in the water supply is an indication of serious contamination with industrial wastes. The maximum contaminant levels for arsenic, cadmium, lead and mercury are shown in Table 30.

The standard water treatment processes are not totally effective in removing such chemical hazards. As the level of heavy metals increases, the plants must obtain other sources of water which have minimum chemical pollutants, or must seek methods of water treatment which will remove that pollution.

Heavy metals are not harmful in very small concentrations, but when consumed by man are not readily metabolized and simply accumulate to a concentration which is toxic.

Only the atomic absorption spectrophotometric methods (AAS-Methods) can be recommended for the determination of such heavy metals. They are accurate and extremely sensitive. Because of the high price of such equipment the analysis can be done in specialized laboratories.

The methods are described in great detail in AOAC [8].

6.3 Microbiological Examination

Although pasteurized and sterilized fruit products receive thermal processes more than sufficient to destroy all pathogenic (disease causing) and spoilage-causing microorganisms, nevertheless, only potable water should be used in preparation and processing of tropical fruit products.

According to the International Standards, safe drinking water should have less than 100 Total Bacterial Count/ml (TBC/ml). Coliform bacteria should be absent from 100 ml of such water.

The results of a TBC of the water supply may be of value in assessing the suitability of water for certain fruit processes, as high count may indicate the possibility that bacteria which may cause spoilage of fruit products are present. Chlorination in the doses usually used in water treatment, destroys the majority of vegetative forms, but has less effort on spore-forming types.

The recommended TBC for water used in the preparation of fruit products is 20/ml, or less.

More dangerous are pathogenic bacteria which can live in water and cause diseases. Of great importance are the following:

Eberthella typhosa (thypoid fever),
Entamoeba histolytica (amoebic dysentery),
Salmonella paratyphi (Paratyphoid A),
Salmonella schottmuelleri (Paratyphoid B),
Shigella dysenteriae (Bacillary dysentery), and
Vibrio cholerae (Asiatic cholera).

For routine control purposes, the direct search for the presence of such pathogenic bacteria is *im*practicable.

Therefore, attention is mainly paid to bacteria species of known excremental origin (faecal matter of warm blooded animals), in particular:

Escherichia coli (and other member of the coliform group), faecal Streptococci *(Streptococcus faecalis)*, and Clostridium welchii.

If such bacteria are present in water, sewage contamination should be suspected, and can be taken as indicator for pollution. They do not multiply in water, but survive for longer periods than the other pathogenic bacteria mentioned before. Hence the coliform count and the presence of Escherichia coli is carried out to determine the presence of faecal contamintion and possible contamination with other pathogenic microorganisms.

Sanitation and treatment which controls the coliform bacteria is adequate for a safe water supply.

Tests of faecal Streptococci and Clostridium welchii are carried out in some cases to obtain additional evidence of faecal contamination and its duration. The chief value of the Clostridium welchii-test is an indicator of the possibility of remove pollution, as the spores are capable of surviving in water for a longer time than organisms of the coliform group.

Municipal water supplies are safe and free of coli contamination. Ground water almost always requires purification to meet drinking water standards. In recent years "in-plant"-chlorination has been widely practices as an aid toward better plant sanitation (see Chapter 7. Manual for sanitation control). Treatment with ozone and ultraviolet can replace chlorination and are highly recommended. The only disadvantage is their being more expensive.

The sterilization is achieved by using high intensity UV lamps which specifically produce the effective germicidal wavelengths between 220–280 nm. Flow rates up to 300 m^3/h are treated by a single arc tube placed in a protective quartz sleeve mounted axially within the radiation chamber.

Ozone is a powerful oxidizing agent. Only a few mg/l in water are enough to destroy microorganisms and to eliminate oxygen-receptive inorganics such as soluble ferrous compounds. The action is more rapid and thorough than chlorine and ozone disappears from water leaving no aftertaste, odour or residual.

6.3.1 Total Count

Determination of the total bacterial count (TBC) in water samples is a common-place and simple examination, is carried out by mixing 1 ml of water with a suitable sterile nutrient agar medium and incubation until the formation of visible colonies which may be counted. Nevertheless, it is recommended that you use the standard plate count (SPC) described by the American Public Health Association, Washington "Standard Methods for the Examination of Water and Wastewater". Especially for comparative and legal purposes [97] the SPC-procedure provides a standardized means of determining the density of aerobic and facultative anaerobic heterotrophic bacteria in water.

Sampling

The sample should be collected without extraneous contamination and analyzed as soon as possible to minimize changes in the bacterial population. Collect the samples in clean sterilized bottles made of glass and provided with ground glass stoppers with an overlapping rim. Allow the water to run from the tap for at least two minutes to flush the interior of the nozzle and to discharge the stagnant water in the service pipe. The recommended maximum elapsed time between collection and examination of unrefrigerated samples is 8 h. Samples stored at less than 10 C should be analysed within 24 h.

Sample Dilution

Buffered water: to prepare stock phosphate buffer solution, dissolve 34 g potassium dihydrogen phosphate, KH_2PO_4 in 500 ml distilled water, adjust to pH 7.2 with 1 N NaOH, and dilute to 1 l.

Add 1.25 ml stock phoshate buffer solution and 5.0 ml magnesium sulphate (50 g $MgSO_4 . 7 H_2O/l$ distilled water) to 1 l distilled water. Dispense in amounts that will provide 99 ml and/or 9 ml after autoclaving for 15 min.

Select the dilutions so that the total number of colonies on a plate will be between 30 and 300.

For most potable water samples, plates suitable for counting will be obtained by plating 1 ml of undiluted sample and 1 ml of sample diluted 1:10 and 1:100.

Plate count agar (Tryptone glucose yeast agar)

Dissolve Tryptone	5.0 g,
– Yeast extract	2.5 g (or 3.0 g Beefextract)
– Glucose	1.0 g, and
– Agar	15.0 g

in distilled water up to 1 l. Adjust the pH to 7.0. Add portions of 15 ml to the tubes, plug the tubes and autoclave at 15 psi for 20 min.

Procedure

Melt the sterile solid agar medium in boiling water and temper the melted medium in a water bath at 45 C until used. Shake the sample and introduce 1 ml into the sterile Petri dish. Add not less than 10 ml of melted nutrient agar. Mix the medium with the sample in the dishes by a circular movement of the dishes, keeping them flat on the table. Incubate at 35 C for 48 h. Check the sterility of the medium, the dilution water blanks, plates and pipettes by pouring control plates for each series of samples. Count all colonies on selected plates promptly after the incubation period. Use an approved counting aid, such as the Quebec colony counter, for manual counting. Automatic plate counting instruments are now available. Compute the bacterial count per milliliter = SPC/ml.

Count each spreader as one colony. Discard plates with spreaders covering half of the plate or more. In such cases, where all the plate counts for the dilutions inoculated are over 300, a fair approximation may be obtained by counting the colonies in a few sectors and calculating the total.

A properly filtered and chlorinated water supply will not show a total count exceeding 10 per ml. The presence of a variety of colonies of different sizes and shapes, presenting a disagreeable appearance and unpleasant odour, indicates undesirable pollution.

6.3.2 Coliform Tests

The plate count is carried out for determining the general bacterial purity of the water and the coliform count for the presence of faecal contamination and possible contamination with other pathogenic bacteria.

The coliform group comprises all of the aerobic and facultative anaerobic, gram-negative, nonspore-forming, rod-shaped bacteria that ferment lactose with gas formation within 48 h at 35 C.

The standard test for the coliform group is carried out by the multiple-tube fermentation technic and involving three tests:

– Presumptive test,
– confirmed test, and
– completed test.

6.3.2.1 Presumptive Test

E. coli is one of the few bacteria which is able to ferment lactose with the production of acid and gas. The absence of gas formation at the end of 48 h of incubation constitutes a negative test. If gas is produced this becomes "presumptive" evidence of faecal pollution. Several non-intestinal bacteria will also produce gas, therefore, a presumptive test must be "confirmed".

In the multiple-tube fermentation technic, the accuracy of test depend on the number of tubes used. At least five tubes should be used.

Materials and culture media: Lactose Broth media (Beef extract 3 g, Peptone 5 g, and Lactose 5 g) is prepared up to 1 l with distilled water. pH should be 6.9. after sterilization. Before sterilization, transfer portions of 10 ml to fermentation tubes containing Durham tubes and sterilize at 15 psig for 15 min. *Sterile pipettes* of 1.0 and 10 ml.

Procedure: Transfer 1 ml portions of the water samples to at least five test tubes containing 10 ml of lactose broth and inverted Durham tubes. Using a 10 ml pipette, transfer 10 ml portions of water samples to five tubes containing double strength lactose broth (26 g lactose broth instead of 13 g) of 10 ml volume. Incubate at 35 C and examine after 24 and 48 h. Formation within 48 h of gas in any amount in the inner fermentation tubes constitutes a positive "Presumptive Test". Gas in all tubes containing 1 ml of water sample indicates gross contamination. On the other hand, a sample showing only one positive 10 ml sample tube should be classed as questionable.

If a quick determination of the presence or absence of Escherichia coli is required the *MacConkey Broth* is recommended as a presumptive test. This broth contains lactose which, when degraded to give acid and gas, indicates the presence of coliform. The gas is collected in Durham tubes and acid production is detected with bromocresol purple. Oxbile promotes the growth of several species of intestinal bacteria and inhibits that of other bacteria. The MacConkey Broth consists of (g/l) Peptone from casein 20 g, Lactose 10 g, Oxbile 5 g, and Bromocresol purple 0.01 g, pH 7.1. The inoculated broth is incubated for 48 h at 35 C.

6.3.2.2 Confirmed Test

Materials and culture media: Tubes of *brilliant green bile broth* with inverted Durham tubes. The broth is prepared by mixing 10 g Peptone, 10 g Lactose, 20 g Oxgall and Brilliant green 0.0133 g up to 1 l with distilled water. pH should be 7.2 after sterilization. Before sterilization, dispense in fermentation tubes and sterilize at 10 psig for 15 min.

Procedure: Gently shake primary fermentation tube showing gas and with a sterile metal loop (3 mm in diameter) transfer one loop full of medium to a fermentation tube containing brilliant green lactose bile broth. Incubate the tubes for 48 h at 35 C. The formation of gas in any amount in the inverted vial of the brilliant green lactose bile broth fermentation tube at any time within 48 h constitutes a positive "Confirmed Test".

6.3.2.3 Completed Test

This test is used as the next step following the Confirmed Test. It is applied to the brilliant green lactose bile broth fermentation tubes showing gas in the Confirmed Test.

Materials and culture media:

Pour plates of sterile *Eosin-Methylene Blue agar* (EMB agar), Levine's modification. to prepare the media, dissolve:

Peptone, 10 g,
Lactose, 10 g,
Dipotassium hydrogen phosphate, K_2HPO_4, 2 g,
Agar 15 g
Eosin Y 0.4 g, and
Methylene blue 0.065 g in 1 l of distilled water.
pH should be 7.1 after sterilization.

Procedure:

Select a positive tube, preferably the highest dilution showing gas or a tube showing gas in 24 h or less. Using a straight inoculating needle, flame and then insert into the liquid in the tube to a depth of 5 cm. Streak the plate by bringing only the curved section of the needle in contact with the EMB-agar surface so that the latter will not be scratched or torn. Incubate the plates at 35 C for 24 h. Typical E. coli colonies will have dark to black centers, button-like in appearance, and will often be surrounded by a greenish metallic shine.

The colonies developing on EMB agar may be described as:

"Typical", nucleated, with or without metallic sheen,
"Atypical", opaque, unnucleated, mucoid, pink after 24 h incubation, or
"Negative", all others.

From each of these plates fish one or more typical well-isolated coliform colonies, tranferring each to a lactose broth fermentation tube and to a nutrient agar slant.

The agar slants and secondary broth tubes are incubated at 35 C for 24 and 48 h if gas is not produces in 24 h. Gram-stained preparations from those agar slab cultures corresponding to the secondary lactose broth tubes that show gas are examined microscopically.

The formation of gas in the secondary lactose broth tube and the demonstration of gram-negative nonspore-forming rod-shaped bacteria in the agar culture may be considered a satifactory "Complete Test", demonstrating the presence of a member of the coliform group in the volume of sample examined.

6.3.3 Coliform Test by the Membrane Filter Technique

The widespread use of the membrane filter technique for a coliform count has confirmed its value, especially its degree of reproducibility, the possibility of testing relatively large volumes of sample, and its ability to yield definite results more rapidly than the standard tube procedure. The membrane filter technique has also been shown to be extremely useful in the routine analysis of water. In a fruit processing plant the use of such a technique may not be necessary.

The membrane technique is used to filter bacteria from water. The trapped bacteria are then grown on the filter by placing it on suitable growth medium. By the use of a selective medium, the number of coliforms and certain other bacterial genera (such as Salmonella, Shigella, and Enterobacter) in the sample can be determined. The *Eosin Methylene-Blue agar* (EMB agar) after Levine, is widely used. The dyes contained in this medium inhibit the growth of many accompanying Gram-positive bacteria. Escherichia coli appears as pink colonies surrounded by turbid precipitation zones. The preparation of this medium was described previously.

The use of *MacConkey Agar* as selective medium is also recommended. Bile salts and crystal violet largely inhibit the growth of the gram-positive microbial flora. Lactose and the pH indicator neutral red are used to detect lactose degradation. The composition of this medium (g/l) is: Peptone from casein 17 g, Peptone from meat 3 g, Sodium chloride 5 g, Lactose 10 g, bile salt mixture 1.5 g, neutral red 0.03 g, crystal violet 0.001 g, and agar 13.5 g. pH value of the medium is 7.1.

After inoculation the plates are incubated for 24 h at 35 C. Lactose-negative colonies are colourless, lactose-positive colonies are red and surrounded by a turbid zone which is due to the precipitation of bile acids as a result of the fall in pH.

Some companies producing equipment and media for making a coliform count by membrane filter technique are:

Millipore Intertech
Sartorius GmbH
Merck, E.
Oxoid GmbH

7   Sanitation Control

The maintenance of sanitation in a tropical fruit processing plant requires that it must be kept clean continuously and not just cleaned at the end of each shift. Sanitation is the maintenance of the work and product environment to prevent hazards of product contamination with food-borne diseases, food poisoning- and spoilage microorganisms, and to provide clean, and safe working conditions.

The term "Sanitation" carries different implications in various regions of the world. The term "Food Sanitation" describes measures taken to promote the safety and quality of foods in North America: Whereas, the United Kingdom describe these measures as "Food Hygiene", because "Sanitation" is associated mainly with sewage and waste treatment.

7.1   Definition and Terminology [106–113]

The following definitions obtained from various sources have been used in texts dealing with cleaning, disinfection, and hygiene of food processing plants. The list is by no means complete but is intended as a guide to preferred terminology.

- *Adjuvant:* A compound that aids, facilitates, or enhances the function of other substance.
- *Adulterate:* To corrupt, debase, or make impure by addition of a foreign, inferior, or harmful substance.
- *Antisepsis:* The destruction or inhibition of microorganisms on living tissues having the effect of limiting or preventing the harmful results of infection. It is not a synonym of disinfection.
- *Antiseptic:* A substance that destroys or inhibits the action of microorganisms on living tissues (as in human body) having the effect of limiting or preventing the harmful results of infection. It is not a synonym for disinfection.
- *Bactericide:* A chemical substance which under defined conditions is capable of killing some bacteria (but not necessarily bacterial spores).
- *Bacteriostasis:* A state in which multiplication of bacterial populations is inhibited. It means inhibition of growth, but not killing, of bacteria by chemicals.
- *Bacteriostat:* A chemical agent which under defined conditions induces bacteriostasis.

- *Break Point Chlorination:* The addition to water of more than sufficient chlorine to satisfy completely the total chlorine demand, and eliminates chloramines or other combined available chlorine residuals.
- *Buffering action:* Stabilizing the pH value of a solution in use. A state in which the hydrogen ion concentration, and hence the acidity and alkalinity, is practically unchanged by dilution, and there is a resistance to change of pH on addition of acid or alkali.
- *Chlorine demand:* Chlorine demand of water is the difference between the dosage level and the residual chlorine which is measured after a contact time of 15 or 20 min.
- *Chelating power:* The property which enables a cleaning solution to redissolve precipitated calcium or magnesium salts.
- *CIP:* Cleaning-in-Place. All of the equipment in a closed system can be cleaned and sanitized without taking the equipment apart.
- *Cleaning:* The removal of food residues, dirt or other objectionable matter.
- *Cleaning agent (or cleaner):* A preparation for cleaning, or an implement or machine for cleaning.
- *Contaminate:* To add foreign and unwanted matter to any object or environment.
- *Corrosion inhibitors:* Substance capable under defined conditions of minimizing the corrosion of certain metals.
- *Detergent:* Substance capable of assisting cleaning when added to water.
- Disinfect: To remove potentially pathogenic (infectious) microorganisms from an object or an environment. Most of bacteria are killed but some spores may survive the process.
- *Disinfection:* The destruction of microorganisms by chemical or physical means.
- *Disinfectant:* A chemical or physical agent that kills pathogenic microorganisms.
- *Dispersing and suspending power:* The ability to bring into and keep in suspension undissolved soiling matter.
- *Emulsifier:* A surface active agent that promotes the dispersion of small fat globules in water.
- *Emulsifying power:* The ability to disperse residual oils and fats and maintain them in suspension.
- *Food hygiene:* A component of sanitary science that is concerned with the preparation, processing, packing, transport, storage or exposure for sale of food so as to ensure that the food is kept clean and fit for human consumption.
- *Fungicide:* A chemical agent which will kill fungi and their spores.
- *Fungistasis:* Inhibition of the growth or multiplication of fungi.
- *Fungistat:* A chemical agent which under defined conditions induces fungistasis.
- *Germicide*: A chemical agent which kills certain microorganisms.
- *Hard water:* Water having alkaline metal ions (mainly calcium and magnesium) at levels more than 60 ppm $CaCO_3$.

- *Hygiene:* The science of the establishment and maintenance of health.
- *Infection:* The condition produced when microorganisms (pathogens) invade tissues and multiply within those tissues.
- *Infestation:* The presence and multiplication of unwanted living organisms in any location.
- *Insecticide:* A chemical agent used to kill insects.
- *Marginal chlorination:* The addition of sufficient chlorine to water to satisfy partially the total chlorine demand of the water being treated.
- *Organic dissolving power:* The ability to solubilize proteins and fats.
- *Pasteurization:* Heating foods to a temperature which will kill pathogenic organisms.
- *Pathogen:* Any microorganism or substance which is capable of producing disease when it enters the human body.
- *Pesticide:* A chemical agent which will kill some kind of living organisms that constitute a nuisance humans.
- *Pollution:* The contamination of an environment with any foreign, unwanted matter.
- *Rinsing power:* The property which ensures that the deposits and solutions can be rinsed from the plant.
- *Sanitary:* Adequately hygienic to ensure a safe, sound, wholesome product fit for human consumption.
- *Sanitation:* The use of practices which will make an environment or substance harmless to human health and well-being. It is the promotion of hygiene and preventation of disease by maintaining sanitary conditions.
- Sanitizing: Treatment by heat or chemicals to reduce the number of microorganisms present, to a level consistent with acceptable quality control and hygiene standards.
- *Sanitizing agent:* (Sanitizer) A chemical agent used for reducing the number of microorganisms to an acceptable level on the product contact surfaces.
- *Sequestering agent:* A chemical agent capable of combining with salts occuring in hard water (calcium and magnesium salts), to form water-soluble compounds and therefore enhancing the detergent operation.
- *Sterilization:* A process intended to destroy or remove all living organisms.
- *Sterilizing agent:* A substance capable of destroying all microorganisms.
- *Surface-active agents:* Substances capable of modifying the physical force existing at surfaces, such as between liquids and solids, permitting more intimate contact and facilitating mixing.
- *Suspending power:* See dispersing power.
- *Swab test:* A direct determination of microbial contamination of a surface.
- *Wetting power:* The property which enables the components to reduce the surface tension of the solutions and thus promote penetration of the soils.

7.2 Factors Affecting Cleaning Efficiency and Costs [109–112]

The selection of cleaning procedures and cleaning conditions to obtain the best cleaning at least cost is a task for both production and quality assurance managers. Optimization for cost-effective sanitation is complex and is a subject demanding detailed in-plant study because of the complex relationship existing between detergent type, concentration, temperature, time, flow rate, soil loads, etc.

7.2.1 Nature of Soil and Soil Formation

Soil in a tropical fruit processing plant is usually easy to remove. On the other hand soil films can harden and set as they dry, making subsequent removal extremely difficult. Caramelization products are more difficult to clean. These deposits can harbour microorganisms which multiply and can subsequently be released into the process stream, along with flakes of scale, by turbulence or thermal cracking. The mechanisms of soil adhesion or adsorption are complex. Soil can be held to the surface by occlusion in surface irregularities, by electrostatic forces between surface and soil or by the attraction of soil fractions such as minerals and protein to each other. During the heat treatment of fruit juices the elevated temperature produces a condition in the juice in which some substances (such as starch, pectin, protein etc.) are no longer in true solution but are in such a state that they either agglomerate or adhere to a surface. Cleaning recommendations always emphasize that surfaces should be cleaned as soon after soiling as possible. Unheated sugar and fruit soils can be easily cleaned provided they have not dried. The ageing phenomenon (time of contact between soil and surface) appears to be of general occurrence. The problem of "fouling" in pipelines, evaporators, and in heat transfer equipment can be avoided by prompt cleaning. Because the surface of a heat exchanger is necessarily hotter than the juice the deposited solids undergo further changes after deposition, usually resulting in a strong soil-surface bond and burning-on occurs.

Strong alkalis are the best detergents for removing burned-on or dried-on materials and for tenaciously adhering deposits.

7.2.2 Water Quality

Water is the universal cleaning tool and is the primary constituent of all food plant cleaners. The water should be good quality, potable and having acceptable bacteriological standards (see Chapter 6. Manual for water control). Water impurities which must be considered with respect to cleaning are:

– *Water hardness* presents the major problem in the use of cleaners by reducing effectiveness and by forming surface deposits. In nearly all cases, detergent costs will increase in proportion to the mineral content, because expensive detergent ingredients will be required to combat water hardness. The water hardness must be known before detailed detergent formulations can be recommended. Soft water (containing less than 50 ppm $CaCO_3$) is preferred for cleaning processes involving alkaline detergents. It gives greater detergent economy and more efficient detergent action, and it prevents scale formation.
– *Soluble iron and manganese* salts should be less than 0.3 ppm to avoid coloured deposits.
– *Suspended matter* must be kept to a minimum to avoid deposits on clean equipment surfaces.

Suspended matter and soluble iron and manganese can be removed only by treatment, whereas water hardness can be partly eliminated by sequestering agents in the cleaning compounds. However, it is more economical to pretreat hard water.

7.2.3 Temperature

The cleaning process can be improved by increasing the temperature. The recommended temperature range is from 40 to 70 C. The rate of soil removal increases by a factor of about 1.5 for every 10 C rise in temperature. The minimum effective temperature, depends on the presence of fat soil, and will be about 35 C, below which fat remains in the solid state. Above 70 C heat-induced interactions and denaturation bind the protein more tightly to the surface and decrease cleaning efficiency. High cleaning temperature can also encourage the burning-on of residual soils.

Increasing the temperature has the following desired effects:

– increasing the solubility of soil materials,
– increasing the chemical ration rates,
– increasing the turbulent action due to decreasing the viscosity of the solution,
– increasing the cleaning process due to decreasing strength of bonds between the soil and the surface.

Hot, or boiling, water is used usually after cleaning of processing equipment, such as plate heat exchangers and pipelines. This means the use of water as a sterilizing/sanitizing aid. Common time and temperature are 85 C for 15 minutes or 80 C for 20 minutes. A higher temperature should be used if the time is to be reduced. The process will not destroy all bacterial spores but is effective against vegetative bacteria, yeast and moulds. Circulation of water at such temperatures is a convenient and effective method for treatment of plant after cleaning-in-place (CIP). The use of steam is also possible. The steaming should be continued for at least 10 minutes after the condensate issuing from the outlet of the equipment reaches 85 C.

A second-choice alternative, to be used if the heating step is not practicable, is to apply a chemical disinfectant after the equipment is visually clean. If any soil remains after cleaning, the cleaning step should be repeated, because virtually all disinfectants suitable for use in fruit processing plants are rapidly inactivated by organic matter. Disinfectants are most effective when applied in hot water. At a temperature of 70 C, a detergent solution kill most spoilage and pathogenic bacteria.

7.2.4 Turbulence

It is obvious, that soil contamination cannot be removed by solubilization alone. Therefore, the use of mechanical cleaning aids is highly desirable to reduce the time and the amount of detergent and to increase the efficiency of the cleanup. Brushes, brooms, water hoses, high-pressure high-temperature water units and steam guns are essential for the continuous cleaning operation.

In clean-in-place (CIP)-systems, the flow of the fluid is utilized in the application of force. The greater, and hence the more turbulent, the flow of detergent solution over the soiling film, the greater will be the chances of penetration, solution, and emulsification. This means, the energy is applied by friction between the deposited soil and the fluid flowing past it. The shear forces generated at a surface are related to the turbulence of the cleaning solution. The change from laminar flow to turbulent flow condition occurs at certain critical flow rates. Successful cleaning is usually obtained only when turbulent flow is achieved.

In practice, the minimum flow velocity required to clean pipelines is 1.5 m s^{-1}. Higher velocities, up to 3 m s^{-1} are required if dead ends, such as T-pieces, are present. In the cleaning of plate heat exchanger used for fruit juices, there appears to be a critical flow velocity (about 5 m s^{-1}) below which cleaning is a function of time of contact, but above which a rapid removal of soil takes place for similar contact times. For this particular cleaning situation, an increase in flow velocity above the critical also leads to a better removal of soils at constant contact time.

To clean moving conveyor belts, pieces of equipments, interior surfaces of tanks ... etc. cleaning by fixed or revolving pressure jets is used. This method ensure that all surfaces are subjected to an effective pressure spray or rinse. The pressure used can vary from 10 to 90 kg cm^{-2}. A number of devices have been developed to achieve a total wetting of the whole surface, ranging from the simple sprayball to the rotating jet system designed not only wet the surface but to give more impingement force between the liquid and surface. The rotating jets are usually much more expensive than sprayballs, but they have a higher impingement factor which allows lower detergent strength to be used.

7.2.5 Time

Soil removal is usually not a spontaneous process. The mechanisms involve a finite time. After a limited time a state where soil redeposition occurs as rapidly as soil removal may be attained. Therefore, increasing the time beyond a given value provides little additional increase in effectiveness. Time must be considered in relation to other variables such as flow rate, temperature and concentration to determine a minimum time for effective cleaning and a practical maximum time for achieving desired results economically.

7.2.6 Nature of Surfaces and Surface Finishes

A detergent should be used which will not corrode or in any way damage the surface being cleaned. Suitable stainless steel is generally considered the surface most resistant to attack by detergents. However, detergents containing hydrochloric acid should be avoided. Mildly acid detergents can be used for cleaning most metals if the contact time is short.

Table 33 shows the composition of *stainless steels* that are suitable for food processing plants. However, for tropical fruit processing plants, the 316 stainless steel is recommended rather than 302 or 304. The basic type 302 is the least costly of any of the standard types mentioned, and is recommended where appearance only is desired and the product does not come into contact with food material. Type 304 is also an austenitic chromium nickel stainless steel like type 302, but containing less carbon and more chromium. It is heat resisting steel somewhat superior to type 302 in corrosion resistance.

Type 316 (AISI 316) is an austenitic chromium nickel stainless and heat resisting steel with more superior corrosion resistance than other types when exposed to many types of chemical corrosives, as well as fruit products. It contains molybdenum. Type 316 also has superior creep strength at elevated temperatures. A very important aspect of stainless steel is its good adaptability to welding, drawing and finishing. The corrosion resistance of stainless steel is due to the formation of a chromium oxide layer which forms naturally on the surface when it is exposed to air. This can be developed artificially by subjecting the surface to hot nitric acid 20–30% solution at 70 C for 30 min and removing the acid. A highly polished surface is also more resistant to corrosion.

Obviously a very rough surface will retain scale and may be difficult to clean. In the food industry for many years a highly polished stainless steel surface has been preferred but such polishing is expensive and must be carried out largely as a hand operation. Electropolishing is widely used and has the advantage of polishing which merely removes the hilltops or, worse still, folds the hills into the valleys, leaving quite definite crevices.

The most popular finish is the so-called Number 4. It is very smooth, easy to clean stainless steel. Following initial grinding with coarser abrasives, sheets are generally finished last with abrasives approximately 120 to 150 mesh.

No. 1 Finish is an unpolished finish, hot-rolled, annealed and descaled stainless steel. It is generally used where smoothness of finish is not of particular importance.

On the other hand No. 7 Finish, which is a mirror finish, has a high degree of reflectivity. It is produced by buffing of the finely ground surface. It is chiefly used for architectural and ornamental purposes. No. 4 Finish is actually as easy or easier to clean than No. 7 and is less susceptible to water spotting.

Conflicting opinions exist regarding the effect of surface finish on soil removal. Some imply that various surface finishes of stainless steel exerted no effect on cleanability as measured by bacteriological tests (swab tests). While recent research indicate that the actual value of surface roughness, quoted as μm Ra, is not necessarily important, but rather the shape of the irregularities determines the clean-ability. The main measuring parameter for surface roughness which is recognized internationally is the arithmetic average of surface deviations above and below the mean surface centre line, Ra, and is identical to that known previously as CLA (centre-line-average). The Ra value is a quantifying concept and bears no physical resemblance to the profile. It can give an indication of the amplitude of surface irregularities, but it takes no accout of their spacing.

Thus, although electropolishing will show a higher surface roughness than mechanically polished steel, there is usually no noticeable difference in cleaning.

On the other hand, other research results indicate that it takes twice as long as to clean a surface of 1.0 μm Ra as it does one of 0.5 μm. It is reiterated that this was a comparative test, and it does not necessarily follow that surface having high Ra values (surface finish more than 1.0 μm Ra) cannot be cleaned with stronger detergent solutions or over longer periods; but the results clearly indicate that the rougher the surface finish, the more difficult it is to clean. Surface finish of about 0.5 μm Ra can be recommended.

Although it would appear that the surface finish of stainless steel does have some small effect on the cleanability of the surface, other factors such as temperature and turbulence play a major part in the cleaning process.

Table 33. Composition of some stainless steels [106]

Element	Type 302	Type 304	Type 316
Carbon	0.08–0.20	0.08	0.10
Manganese	2.00	2.00	2.00
Phosphorus	0.04	0.04	0.04
Sulphur	0.03	0.03	0.03
Silicon	1.00	1.00	1.00
Nickel	8.00–10.00	8.00–10.00	10.00–14.00
Chromium	17.00–19.00	18.00–20.00	16.00–18.00
Molybdenum	0.00	0.00	2.00–3.00

American Iron and Steel Institute, New York, USA.

Aluminium is widely used in some countries for the construction of fruit processing equipment, because of its low cost, light weight and ease of fabrication. The principal objection to the use of aluminium has been its lack of resistance to abraison and susceptibility to corrosion from cleaning solutions.

Aluminium is resistant to corrosion from atmospheric condition, to water and to fruit juices, but is particularly attacked by caustic solutions. A plant with one piece of aluminium equipment makes it necessary to use an aluminium cleaner throughout the plant, which can be somewhat expensive. Moderate alkalis with high metasilicate content are suggested for cleaning these surfaces. For washing cans and preventing scales on coolers, a detergent containing trisodium phosphate is suggested. Sodium metasilicate may also be used for washing cans without danger of corrosion.

Glass surfaces are etched by strong alkalis and should be cleaned with moderately alkaline. Detergents high in polyphosphates are recommended.

Rubber surfaces are not affected by alkalis and may be cleaned with any alkaline detergent. However, strong acids and organic solvents should not be used on rubber.

Concrete floors are etched by acids, but are not harmed by alkalis. Detergents high in metasilicate are suggested for cleaning concrete floors.

7.2.7 Detergent Formulation and Concentration

Detergents can be defined as substances capable of assisting cleaning when added to water. They include soaps, organic surface active agents, alkaline materials and acids in certain instances. In the tropical fruit processing industry the predominant soil is sugar and various fruit cells, no real fatty residues or heavy protein depositions are present. Therefore, it is becoming increasingly popular to blend one's own specific chemicals from a simple range of basic chemical products. This enables the user to prepare an infinite variety of concentrations and formulations for his own specific requirements.

Cleaning compound selection depends upon a number of inter-related factors which include:

– type and amount of soil
– nature of surface,
– water quality,
– method of cleaning available, and
– costs of detergents.

Most of these factors have been discussed previously. The cleaning methods available and costs of available detergents cannot be overemphasized in the cleaning process. Where cleaning is done by hand, it is evident, that strong acids and alkalis cannot be used. Therefore, superior results can be achieved by use of circulation cleaning, in which the optimum concentration of cleaning compounds can be more readily utilized.

The true value of a detergent is expressed in terms of the unit of surface that can be efficiently cleaned at a minimum cost. Frequently, high-cost substances (cost in US $ per kg) are most economical because higher cleaning efficiency saves time and external energy requirements and reduced compound requirements.

Because selection of the proper combination of cleaning agents and procedures can be rather complicated, many plants prefer to buy commercially formulated products which are sold under trade names at prices substantially above the cost of the ingredients. However, the suppliers usually provide their customers with services and advice about water hardness, pH adjustment, soil characteristics, surface corrosion, temperature and other factors that effect the cleaning process.

Types, functions, and concentration of detergents used in the fruit processing industries can be illustrated using the following basic categories:

- dissolving action,
- sequestering power,
- wetting power,
- dispersive and suspensive power,
- buffering action, and
- bactericidal power.

Dissolving Action

The dissolving action depends mainly on the presence of water, which is solvent and carrier for soils, as well as chemical cleaners. For the detergent formulations usually used in fruit processing plants, the dissolving action results from the active alkalinity, quoted as Na_2O. For CIP-Procedures concentrations between 1000 to 3000 ppm are recommended, while for hand cleaning procedure 500 to 800 ppm are satisfactory. Active alkalinities of some detergents are given below:

	% active Na_2O
Strong alkalis	
Sodium hydroxide (caustic soda)	76
Sodium orthosilicate	46
Sodium sesquisilicate	36
Mild alkalis	
Sodium carbonate (soda ash)	29
Sodium metasilicate	28
Sodium sesquicarbonate	
(compound of sodium carbonate and bicarbonate)	13
Trisodium phosphate (water softener)	10
Sodium bicarbonate	0

Strong alkalis are used in concentrations of 0.4 to 3% and mild alkalis of 1 to 5%. *Caustic soda* is the cheapest and strongest alkali. It is corrosive to many surfaces and dangerous to the skin. It has no buffering properties, and precipitates mineral hardness from water. Therefore, caustic soda is seldom used alone as detergent.

Sodium orthosilicate (a compound of sodium hydroxide and silica) has no buffering action and precipitate water hardness, but is better than caustic soda in the other properties such as wetting or penetrating action and rinsing properties.

Sodium metasilicate is mild alkali widely used in fruit processing plants because it is more desirable as a detergent ingredient. It is less corrosive to metals, a good buffer, it has less tendency to precipitate water hardness, has good rinsing and penetrating properties. It is more expensive than sodium hydroxide and sodium carbonate. *Sodium sesquisilicate* is a compound intermediate in properties between the meta- and orthosilicates.

Sodium carbonate (soda ash) is a cheap alkali having similar properties to caustic soda, but is a weaker alkali, less corrosive and acts as a buffer. Because of their low alkalinity, sodium carbonate and sodium bicarbonate are used in detergent formulation which come into contact with the skin.

Trisodium phosphate (TSP) is a more expensive source of alkali than sodium metasilicate but is about equal in most properties. Nevertheless, it was widely used in detergents for equipment and in dishwashing compounds because of its ability to prevent hard water deposits on cleaned surfaces. It does not however, have the sequestering properties of the *polyphosphates*. Generally, the polyphosphates (such as sodium tripolyphosphate) are excellent as detergent ingredients, being: good water softeners, dispersing agents, suspending agents and free-rinsing. However, they are more expensive than trisodium phosphate.

Nowadays phosphates are being used less and less due to environmental problems.

Acids are used to dissolve carbonate scales and certain mineral deposits. For such purposes *inorganic acids* (nitric, phosphoric or sulphamic acids) are used in concentrations of about 0.5%. These materials should be handled with care because they can cause severe burns to the skin and irritate mucous membranes. In addition they are very corrosive to certain metals, and for this reason they may have to be used with corrosion inhibitors. *Organic acids* (acetic, hydroxy-acetic, or lactic acids) are moderately corrosive but can be inhibited by various organic nitrogen compounds.

Sequestering Power

Sequestering is the ability to prevent deposition of undesirable mineral salts on the surfaces being cleaned. There are three main classes of sequestering or chelating agents, which can form soluble complexes with calcium, magnesium and iron to prevent film formation on equipment and utensils:

– Ethylenediamine tetraacetic acid (EDTA and its salts)
– Gluconic acid (and its salts), and

– Polyphosphates (such as sodium tripolyphosphate, hexameta- phosphate, tetraphosphate and pyrophosphate).

The amount of sequestering agent necessary depends on the hardness of the water in which the detergent is to be used. Cost, effectiveness, and stability in hot alkali solution are important in the selection of a sequestering agent. Polyphosphates are cheapest and best for moderately alkaline detergent. They have fair sequestering powers. Sodium triphosphate is the most stable in hot alkaline solution, while sodium hexametaphosphate is the least stable. For this reason such detergents containing polyphosphate should be made up only as needed.

Sodium salts of gluconic acid are stronger chelating agents than EDTA. However, their optimum sequestering action is developed in caustic soda solutions of 2–3% strength, which is far above that normally used in fruit processing plants.

Wetting Power

Wetting is the ability to lower the surface tension of the water medium so as to increase its ability to penetrate soil. Surfactants (surface-active agents) fulfil this requirement and contribute many other useful properties to detergent compounds, namely emulsification, dispersion, suspension, and improved rinsability. Some surfactants possess all these properties whilst other contribute only one or two. Cost and stability of the wetting agent under conditions of use are also important in its selection. Generally the approximate concentration of use is about 0.15% or less.

Wetting agents usually used can be classified as anionic, non-ionic, or cationic depending upon how they dissociate in aqueous solution.

Anionic wetting agents (soaps, sulphated alcohols, sulphated hydrocarbons, aryl-alkyl polyether sulphates, sulphonated amides and alkyl-aryl sulphonates) is the most common group and consists of essentially neutral materials that can be adjusted for use in either acid or alkaline media. They are not compatible with cationic wetting agents and some of them foam excessively. *Alkyl-aryl sulphonates* are the most common, however, the sulphated fatty alcohols may also be used. Both have good wetting properties, but the alkyl-aryl sulphonates are cheaper. Synthetic household detergents are primarily of the sulphated fatty alcohol group.

Non-ionic wetting agents (polyethenoxyethers, ethylene oxide-fatty acid condensates and amine-fatty acid condensates) are excellent detergents for oil, but many of them are liquids. As they do not ionize, they are compatible with either anionic or cationic materials.

Cationic wetting agents (quarternary ammonium compounds) have some wetting effect and are not compatible with anionic wetting agents. *Quarternary ammonium compounds* are dealt with later, as their main use is as antimicrobial agents.

- *Dispersive and suspensive power*
  Suspending agents assist in keeping undissolved soil in suspension. A typical example of a substance that inhibits redeposition of soil is sodium carboxymethyl cellulose.
- *Buffering action*
  Sodium sesquicarbonate, sodium polyphosphate, sodium meta-silicate, and to a lesser extent, sodium carbonate and trisodium phosphate, acts as buffers in detergent solution. Buffering action is desirable to give a reserve effect and maintain the pH.

If the solution is to come into contact with the hands to any extent, it is advisable to have a pH below 11.4.

Bactericidal Power

Some of the alkaline cleaners, such as phosphates, carbonates, silicates and especially sodium hydroxide, greatly facilitate the removal of microorganisms from surfaces and are also lethal to them when applied in strong solutions (5–10%). Their use in bactericidal concentrations is restricted mainly to many surfaces, because they are highly corrosive for a wide range of organic and inorganic materials, including aluminium, human skin and mucous membranes.

A large number of chemical sanitizing agents are available, but only three types are in use in fruit processing plants: chloride bearing compounds, quaternary ammonium compounds and iodine and iodine complexes. As normally used, chemical sterilizing agents are not effective against bacterial spores, and should not be relied on to kill mould spores.

Chemical sterilizing agents may be formulated with detergents to provide balanced products which clean and sterilize in one operation, and at temperatures below 60 C. At temperature of 70 C, detergent solutions alone kill most spoilage and pathogenic bacteria, and the inclusion of a chemical sterilizing agent may thus be unnecessary at this stage.

Sanitizing agents are most active when used after the cleaning process.

7.2.8 Disinfectants (Sanitizers)

In food processing plants, disinfection or destruction of food-spoiling and disease-causing microorganisms can be considered to be virtually synonymous with chlorination. Various other disinfection methods such as heating, irradiation with ultraviolet rays, and the addition of ozone, iodine or alcohol have been advocated from time to time, and some of these are being used to a limited extent but no other technique has proved to be more effective than, or as inexpensive as chlorination.

It seems appropriate to note that heat in the form of

- steam (80 C for 15 min, or 95 C for 5 min),
- hot water (80 C for 5 min), or
- hot air (80 C for 20 min)

is the most recommended and most effective means of destroying microorganisms in fruit processing plants.

Ultraviolet radiation has a major application in packaging areas. Contact time should exceed 2 min, and this destroys only those microorganisms that are in direct rays of the light. As mentioned before, some of the alkaline and acid cleaners, greatly facilitate the removal of microorganisms from surfaces and are also lethal when applied in strong solutions at temperature higher than 70 C. Chemicals that kill microorganisms more or less nonselectively are termed disinfectants when used in medical practice. They are also known as sanitizers when applied elsewhere to utensils, and equipment, as in food processing plants.

Hypochlorites and quaternary ammonium compounds are the leading disinfectants in food sanitation operations.

Quarternary Ammonium Compounds (QACs)

Many forms are available, including:

- Alkyldimethylbenzyl ammonium chloride,
- Methyldodecyl benzyltrimethyl ammonium chloride,
- Lauryldimethylbenzyl ammonium chloride,
- Alkyltolylmethyl trimethylammonium chloride, and
- Didodecenyldimethyl ammonium chloride.

QACs are used in concentrations of 150 to 200 ppm at temperatures above 40 C and with a minimum contact time of 2 minutes. Low concentration, low temperatures, hard water, presence of organic food matter reduce the bacterial activity of QACs.

Unlike hypochlorite, QACs should not be added to anionic wetting agents or acids, because they may be seriously inactivated. On the other hand, increasing the alkalinity, enhances the bactericidal activity of QACs.

Compared with undiluted hypochloride they are much safer to handle and are relatively non-corrosive to metals. QACs are active against many microorganisms, especially the thermoduric types, however, they do not kill bacterial spores but may inhibit their growth.

The main disadvantage of QACs are expensiveness, formation of foam in mechanical applications and the noncompatibility with anionic detergents and hard water.

Chlorine Compounds

Chlorine and chlorine-releasing compounds may be used as sanitizing agents alone or may be added to solutions of suitable detergents to provide combined detergent-sanitizers. Sodium and calcium hypochlorites and chloramines are preferred for most sanitizing operations in food processing plants, because they are safer and easier to use than liquid chlorine. Organic chlorine-release agents, such as dichlorodimethyl hydantoin, are also common in detergent formulation.

Hypochlorite solution, containing 50–200 ppm active chlorine in water, are often used as the final germicidal rinse, at temperature exceeding 40 C for longer than 10 min. For tropical fruit processing plants, rinsing with cold solutions containing 100 ppm for 15 minutes for previously cleaned equipment is adequate. Residual organic matter reduces the activity of chlorine. Storage of used solution may result in a marked decline in strength, and only freshly prepared solutions should be used. Where possible, the available chlorine content of the solution should be checked at the time of use (Sect. 6.2.10). Hypochlorite solutions are corrosive to most metals including stainless steel. Therefore, using low concentrations in alkaline solutions, at low temperatures with short contact time, the sanitizing action remains effective but with a minimum of corrosion hazards.

The advantages of chlorine are inexpensiveness, it is unaffected by hard water salts, is active against all microorganisms and also against spores at high temperatures and with long contact times.

7.2.9 Chlorination of Water

Chlorination of water has become a common practice in fruit processing plants as a means of improving plant sanitation. In general, the maximum content of active chlorine in drinking water must not exceed 0.3 ppm. On the other hand, the cooling water should contain 2–4 ppm of free available chlorine to keep total bacteria count within safe limits. Cooling canals and systems are susceptible to microbial build-up during operation.

Chlorination of water used for cleaning equipment, belts and floors to levels between 10–20 ppm available chlorine is required for killing disease-causing bacteria, slime, algae, iron bacteria and also aids in the precipitation of iron and manganese. Odours developing from organic matter are prevented by such chlorination.

Scale formation can occur in the cooling system, and the addition of a small amount (5 to 15 ppm) of sodium hexametaphosphate will help control scale, inhibition of corrosion can be achieved by adding orthophosphate and chromate to cooling water.

Chlorine is corrosive to most metals. However, 2–4 ppm does not noticeably corrode either cans or equipment under normal conditions. Even 10–20 ppm of chlorine, used for clean-up, produces no significant corrosion because the contact time is too short. Equipment washed with high concentration of chlorine should be rinsed after 15–20 min.

Selecting a Chlorine Source

The chlorine source to be used will depend mainly on the volume of water to be chlorinated. Where a large amount of water is to be chlorinated, chlorine gas is recommended. Though the initial cost is high, it is balanced by the lower cost of the chlorine. Hypochlorites are good sources when only small amounts of chlorine are needed.

Chlorine is a green gas whose odour is perceptible even in concentrations as low as 0.001 mg/l air. The density of chlorine is about 2.5 times that of air. Therefore, chlorination equipment should not be located in a basement as leakage accumulation could reach toxic concentrations (over 50 mg/l air). Precautions should be taken to reduce the possibility of personal content with chlorine gas. Chlorine cylinders and chlorination equipment should be located out of doors at cold temperatures in a well-ventilated and fireproof area with a minimum of personnel traffic.

In fruit processing plants, chlorine is usually applied to water in the form of hypochlorites solutions, either by simple drip addition, or through proportioning pumps. Sodium hypochlorite ($NaClO$) consists of readily water-soluble crystals and is also marketed as an aqueous solution. Equipment costs are lower than that of chlorine gas, but hypochlorites are more expensive. Materials which contain phenols or cresols, such as marking inks, paints, fly sprays, boiler feed water compounds, etc. should be removed from the plant because chlorine on reaction with them, even in minute quantities, produces compounds of undesirable flavours.

Chlorine Demand, Marginal Chlorination and Breakpoint Chlorination

Chlorine Demand of water is the difference between the dosage level and the residual chlorine which is measured after a contact time of 15 or 20 min.

The amount of chlorine consumed by the reaction with impurities present in water depends mainly upon the kind and amount of the impurities and also upon pH, contact time and temperature. The impurities responsible for chlorine demand include compounds containing iron, manganese, nitrites and sulphides. The chlorine which reacts with these compounds has no germicidal properties and cannot be measured by the methods used for testing chlorine concentration. Usually 0.25 to 0.75 ppm of chlorine enters into this reaction.

The amount of chlorine remaining in the water after the demand is satisfied is referred to as total *residual chlorine*. The total residual chlorine includes the *free* and *combined residuals*. Some chlorine will usually react with nitrogenous compounds and will be included in the combined residual.

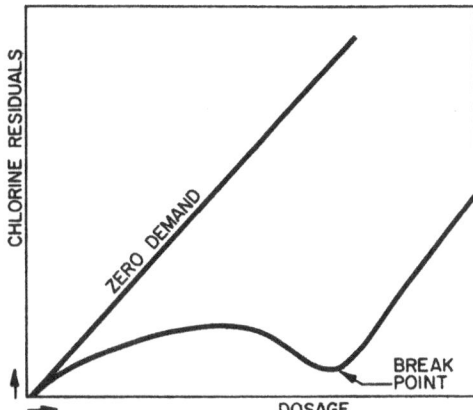

Fig.22. Break point chlorination
of water

Marginal Chlorination means the addition of sufficient chlorine to water to satisfy partially the total chlorine demand of the water being treated. It is attained by adding sufficient chlorine to produce temporary chlorine residuals of somewhat less than 0.5 ppm, measured at a point not far beyond the point of addition. This treatment serves to destroy coli type organisms. However, because the chlorine demand of the water itself is not satisfied, the chlorine residuals produced in marginal treatment are dissipated quickly. There is then no available chlorine in the water.

As chlorine is added to water, a part is utilized in meeting the chlorine demand. In addition, some chlorine combines loosely with nitrogenous matter to form chloro-nitrogen compounds. As a free chlorine residual begins developing in the water some enters into an oxidation reaction with the chlor-nitrogen compounds. The free chlorine residual level drop slightly until the oxidation reaction is complete and then will rise according to the dosage as indicated in Fig. 22. Different water supplies demonstrate different Break-Point Characteristics.

Break Point Chlorination means the addition to water of more than sufficient chlorine to satisfy completely the total chlorine demand, and eliminates chloramines or other combined available chlorine residuals. After this point the free chlorine residual increases in direct proportion to the amount added. Break Point Chlorination usually eliminates the undesirable odours and flavours commonly associated with Marginal chlorination, and supplies persistent free chlorine residuals. For normal in-plant chlorination a residual of 2 to 5 ppm available free chlorine is required. During clean-up the chlorine level is boosted to 10–20 ppm or higher.

Determination of Residual Chlorine: See Sect. 6.2.10.

7.2.10 Cleaning Procedures

The main reason for cleaning the processing line and processing areas is to remove fruit debris and other soil which contain spoilage microorganisms and also to maintain an environment of cleanliness that encourages a high standard of hygiene among personnel.

The basic stages and sequences of cleaning and disinfection given below can only be regarded as guidelines as they will vary from a plant to other:

- Pre-rinse with cold or hot water to remove gross soil. The proper temperature of the water will depend on the type and amount of soil, and the nature of the equipment.
- Detergent wash using 0.5–1.0% caustic base detergent containing wetting and sequestering agents.
- Scrubbing of the soiled surfaces, if necessary.
- Intermediate water rinse, using hot water to remove suspended soil.
- Acid wash, if necessary.
- Final water rinse to remove remaining acid.
- Hot-water sterilization or application of a chemical disinfectant.
- Rinse away the disinfectant, if using potable water.

In order to clean, successfully and economically, a fruit-processing plant and equipment, the initial design of the plant must take into consideration the method of cleaning after processing. There are three different methods of cleaning in a tropical fruit-processing plant:

- Hand cleaning using brushes and abrasives,
- High-pressure cleaning equipment, and
- Clean-in-Place (CIP) systems.

Hand Cleaning

Small equipment, containers, painted surfaces, plastics etc, can be cleaned by hand with brushes or by pressure jets. For difficult cleaning jobs abrasives, such as pumice, silica flour, steel wool, metal or plastic "chore balls" can be used. The disadvantages of such procedures are scratching of surfaces, damage to workers' skin and that particles may become embedded in equipment and later appear in the food.

For manual cleaning, it is convenient to use the following system:

- Pre-rinse with running water to remove bulk soil.
- Wash with a suitable detergent solution. The pH of the solution should be less than 10 to avoid skin irritation and the temperature at about 50 C. Chlorinated alkalis should be avoided because of excessive skin irritation.
- Post-rinse with hot water (70–80 C) for 1 to 2 min.

Parts of equipment can be washed in washing machines or by soaking in detergent solution containing chemical disinfectants, and allowing to remain in contact for several minutes. All cloths, brushes, sponges and other cleaning aids should be washed and disinfected frequently. Washing with hot detergent and soaking in solution containing 200 ppm available chlorine is recommended.

High-Pressure Cleaning Equipment

High pressure water equipments may be either stationary or movable. High pressure cleaning is based on atomization of the cleaning solution through a high pressure spray nozzle. Pressure may be developed with piston type pumps and multistage turbine pumps which may operate at pressures as high as 800–1000 psi. For best results, the water pressure must be over 15 psi and the steam pressure above 50 psi. Cleaning effectiveness is dependent largely upon the force of the cleaning solution against the surface, which is controlled by the nozzle design. The movable units are generally small and are well adapted for small plants or specific areas within large plants. Stationary units are installed with high pressure lines piped to the various departments for cleaning use. High-pressure cleaning is recommended for cleaning the exterior parts of equipment, large free-standing pieces of equipment, tanks and for floors and walls.

The system uses relatively small amounts of water, and provides a convenient means of cleaning crevices and large surfaces. However, it must be used with care, because the spray may contaminate other equipment. The type, size, and working pressure must be chosen to match the cleaning problem.

Clean-in-Place (CIP) Systems

CIP is being increasingly used in modern fruit processing plants for cleaning pipe lines, heat exchangers, properly designed tanks, evaporators, and filling machines.

A CIP system includes a fully controlled efficient cleaning schedule of time, temperature, detergency and of turbulence. CIP systems were originally designed on the basis of installing a centralized plant complete with all necessary pumps and heating equipment; and on the use of detergents only once and then throwing them away.

The current trend is to install a partly centralized system employing self-contained satellite cleaning units. A number of these satellite units can be installed throughout the plant and each unit contains its own control panel. In single-use systems, the detergents are used for cleaning only once, but with the growth and complexity of plants and the number of circuits to be cleaned, the re-use of detergents is adopted.

A single-use system is suitable for very heavy soil loading where contamination of the cleaning fluids is such that recovery is not economic. In the re-use systems, the used solution is filtered, made up to strength and stored ready for the next cleaning cycle. The system is satisfactory for moderate soil loads.

CIP systems are very expensive to design and to install but are much more cost-effective when saving of water, detergents, heat, downtime and labor costs are taken into account. Other advantages of CIP systems are the greater convenience, better hygiene and safety of operators. There is no need to crawl into tanks or come into contact with detergent and sanitizers. Manual cleaning, on the other hand, requires little capital investment but substantially greater labor costs. The high capital cost for CIP systems is because the equipment should be designed so that it can be cleaned in place, or because it requires modification of some equipment for automation.

A typical program for CIP would be as follows:

– Preliminary water rinse to remove gross soil,
– Detergent wash to remove residual soil,
– Intermediate water rinse to remove detergent,
– Acid circulation (if necessary) to remove scales,
– Intermediate water rinse (if necessary) to remove acids,
– Sterilant circulation to destroy residual microorganisms, and
– Final water rinse to remove any CIP solutions.

In conclusion it must be noted that the cleaning and sanitation procedure and application may well be doomed to failure unless adequate staff training is arranged and a planned quality assurance procedure adopted.

For those who want to know more about cleaning, sanitizing, and about hygienic operation of food processing plants several publications are available [32, 50–56].

7.3 Evaluation of Cleanliness and Sanitation

Cleaning and sanitation in a tropical fruit processing plant should be planned maintenance of the work and product environment to prevent hazards of product contamination and conditions aesthetically offensive to the consumer, and to provide clean, healthy and safe working conditions. Evaluation of cleaning and disinfection procedures has been limited by the difficulty of determining the degree of soil removal. Several methods have been recommended for such an evaluation, including electrical conductivity, radioactive tracer techniques, light-scattering, light-transmittance, fluorochromatic techniques ... etc. Such methods are impractical, expensive and usually inaccurate. The universally acceptable

Fig. 23. Plant and equipment cleanliness in a fruit juice concentrate and fruit puree processing plant [EURO CITRUS]

means of measuring cleaning efficiency are visual evaluation and microbiological evaluation.

7.3.1 Visual Evaluation

A clean surface is a surface, free from residual soil and film. The contamination will not be visible under good lighting conditions with the surface wet or dry. Several reliable visible tests are used to determine the degree of cleanliness:

– The surface does not give a greasy or rough feeling to clean fingers when they are rubbed on the surface.
– The surface shows not sign of excessive water break while water is draining from it.
– The droplet test – droplets adhere to unclean surfaces.

- The salt test – salt sprinkled on wet surfaces to make the adhering moisture more visible.
- The carbonated water test – gas bubbles adhere to soil films on unclean surfaces.
- The paper test, – a new, white paper tissue wiped several times over the surface shows no discolouration.
- Ultraviolet light test – no sign of fluorescence should be detected when the surface is inspected with long wave ultraviolet light radiation (340–380 nm).

7.3.2 Microbiological Evaluation

Evaluation of plant, equipment, working surfaces, walls, floors and air cleanliness is based mainly on microbiological analysis, which provides an index of actual sanitation procedures, discovers environmental sources of spoilage organisms in response to shelf life problems and of foodborne pathogens. Microbiological tests may also be carried out to evaluate the design of food processing facilities and equipment in terms of sanitation and to determine the necessary frequency for special maintenance procedures.

Ideally, product contact surfaces of equipment and air should be practically free from microorganisms. On the other hand, bacterial count cannot be recommended as a sole measure of cleanliness. The usual standard required is physical cleanliness and a microbial count of less than 1 organism per square cm. Various procedures and criteria to evaluate equipment and plant cleanliness have been developed. Certain advantages are claimed for each technique, but each has its own limitations. The following methods can be recommended:

- Rinsing test,
- Swab test,
- RODAC-Method, and
- Membrane filtration of air.

Rinsing Test

The rinsing test is an indirect microbial determination made by rinsing with a sterile liquid and subsequent determination of the degree of microbial contamination. The test is suitable for containers, closed systems (CIP-Systems), and piping networks. It can be used as a complete control, if the whole system is rinsed, or as a localized control.

Media and Reagents:

- Ringer rinse solution, dissolve 2.15 g NaCl, 0.075 g potassium chloride, 0.12 g anhydrous calcium chloride and 0.5 g sodium thiosulphate in 1 l of distilled water. Sterilize in an autoclave for 20 min at 1 bar. The ringer

solution is ionically balanced and is used at one-quarter of the original strength. A 0.85% NaCl solution has the same osmotic pressure as microorganisms.
- *Plate count agar,* dissolve 5 g tryptone, 2.5 g yeast extract, 1 g glucose and 15 g agar in 1 l distilled water by boiling, and adjust to pH 7.1. Dispense into the tubes or flasks and autoclave 15 min at 121 C. To make plate count agar with bromocresol purple, add 0.04 g BCP per litre of medium.
- *Sterile pipettes, sterile Petri dishes and 70% ethanol.*

Procedure: Add 20 ml of sterile rinse solution to each container and aseptically recap the container. For bins and large utensils, add 500 ml solution. Shake the container or rotate the vessel for 10 min. Make the count with 1 ml of the rinse solution in duplicate. Pour the plate with 15 ml of the desired medium, e.g., Plate Count Agar or other media. Membrane filtration procedures should be used if low levels of contamination are expected or when 100-ml or greater portions of the rinse solutions are used, such as of rinse solution from CIP processing assemblies. Both numbers and types of microorganisms should be taken into consideration. The types of microorganisms from container rinse samples may be important in terms of their potential to cause spoilage. Under certain circumstances such as with aseptic packaging, the microbiological condition of containers is a critical control point.

For food contact surfaces, the usual standards require that containers should have a residual bacteria count of one colony or less per ml of capacity or not over one colony per square cm of product contact surface. Coliform bacteria should not be present.

Swab Test

The swab test is a direct detection of microbial contamination by transfer from defined surfaces. The test is usually the method of choice. The test is used for large areas for sampling utensils, tableware, kitchenware. Swab procedures should also be used in difficultly accessible parts of the plant and of the equipment such as cracks, corners or crevices. However, the swab test is a qualitative test. No exact quantitative statement is possible.

Materials and Reagents:

- *Cotton wool swabs,* wind non-absorbent cotton wool on wooden sticks or glass rods about 12 to 15 cm long to form a swab of about 0.5 cm in diameter by 2 cm long. Swabs should be packaged in individual or multiple convenient protective containers (e.g. test tubes) with the swab heads away from the closure. Plug with non-absorbent cotton wool and sterilize in a hot-air oven for 90 min at 150 C.
- *Swabs made of calcium alginate,* such swabs are highly recommended. Swabs made of calcium alginate fibers are soluble in aqueous solutions (rinse,

culture media etc.) containing 1% of sodium hexametaphosphate (or sodium glycerophosphate, or sodium citrate, or 1% of any mixture of these). All organisms dislodged from swabbed surfaces are thus liberated.
– *Vials*, small screw-capped, 7 to 10 cm long, prepared to contain 5 ml of rinse solution after autoclaving.
– *Plate count agar, sterile pipettes, sterile Petri dishes*, and 70% ethanol.

Procedure: Open the sterile swab container and remove the swab aseptically. Open a vial of rinse solution, moisten the swab head, and press out the excess solution against the interior wall of the vial with a rotating motion. Rub the swab head slowly and thoroughly over approximately 50 square cm of the surface, three times reversing direction between successive strokes. Return the swab head to the solution vial, rinse briefly in the solution, then press out the excess. Swab for more 50 square cm areas of the surface being sampled as above, rinsing the swab in the solution after each swabbing, and removing the excess.

When unmeasured surface areas (pump impellers, rings and valve seats) have been swabbed, the results may be reported on the basis of the entire sampling site instead of a measured area.

Plate 1 and 0.1 ml portions of rinse solution after shaking the vials manually or by means of mechanical shakers. Pour plates with Plate Count Agar or other appropriate media, depending on the organisms of interest. Incubate plates, count colonies, and calculate the number of colonies recovered from 50 cm^2 (equivalent to 1 ml of rinse). In the case of utensils, report the count as the residual bacterial count per utensil examined.

Bacterical count not exceeding 100/50 cm^2 (averaging 2 colonies per square cm) is satisfactory. For unmeasured surface areas, the level of a few colonies per sampling is usual.

In many cases, the types of microorganisms may be more significant than the numbers alone. For example the presence of even very low numbers of Saccharomyces bailii and/or Lactobacillus spp. in fruit processing equipment may be highly significant with respect to the potential spoilage of fruit products.

RODAC-Method

RODAC (Replicate Organism Direct Agar Contact) provides a simple agar contact technique and is recommended for flat impervious surfaces. It is not suitable for heavily contaminated areas, and also not suitable as a negative control, as the surfaces tested are too small. A sufficient number of spots should be sampled to obtain representative data. Disposable plastic RODAC plates may be purchased prefilled with test medium or they may be filled in the laboratory. Normally, Plate Count Agar is used. They should be incubated at 32 C for 24 h as a sterility check.

Procedure: Remove the plastic cover and carefully press the agar surface to the surface being tested. Replace the cover and incubate in an inverted position for

24 h. Colonies should be counted using a Bactronic or Quebec colony counter and recorded as the number of colonies per RODAC plate, or number of colonies per square cm.

Membrane Filtration of Air

In a tropical fruit processing plant the microbiological quality of air does not directly affect the shelf-life of the product. Only in the areas of packing, especially for aseptically packaged fruit product, the presence of bacteria, bacterial spores, mould and yeasts in the air is highly significant for contamination of soil and dust. When personnel are the source of microbial contamination, the primary types are vegetative bacteria, especially staphylococci, streptococci and micrococci, which are associated with the human respiratory tract, hair and skin. Several methods are available for microbiological monitoring of air. The method of choice is membrane filtration. In this method a standard membrane filter (0.45 μm) incorporated in a sampling housing connected to a vacuum source and flowmeter is used. The microorganisms are trapped on the filter which after the sampling period is incubated on an appropriate culture medium. The bacterial count is calculated per unit volume of air. Examples of such equipment – Millipore membrane filterfield monitor and Sartorius Membrane Filters.

Microbiological monitoring of air using slit samplers is also common. These are agar impaction samplers. They are cylindrical and have a slit tube which is threaded onto the top of the sampler and may be adjustable. Beneath the slit is a platform which accepts a culture plate and which is rotated by a clock mechanism. Air is drawn through the slit impacting suspended particles onto the rotating agar surface. After sampling, the plate is removed and incubated as required and colonies counted to determine the number of viable particles per cubic meter of air over defined period of time.

Some companies producing slit samplers are BGI Incorporated and the New Brunswick Scientific Company.

7.4   Control of Employee Hygienic Practices

The human factor in sanitation is among the most important because the human must carry out sanitation practices and because every human is a potential carrier of disease-producing and food-spoiling microorganisms in and on his body. Understanding, good health and personal hygiene of all workers are, therefore, important factors in reducing the transfer of microorganisms to the food. The quality assurance manager must train employees and supervise them so that a high level of protection is maintained. A food processing plant can be kept in good sanitary condition only so long as the personnel practice good personal hygiene as well as good sanitation. The following aspects are especially important:

Design and Layout of Employees Facilities

Rest rooms, dressing rooms, locker rooms, toilets, wash rooms, hand washing facilities, eating facilities, drinking water fountains etc should be well-maintained for the comfort, hygiene and safety of the employees if they are to remain happy, conduct themselves with pride and care reflected in production efficiency and product quality. It is essential that clean, uncluttered, and adequate space is provided for employee facilities. Hand washing facilities in addition to those in dressing and rest rooms should be provided in areas adjacent to, but not actually in, fruit handling areas. All facilities must be cleaned and disinfected at the same regular and frequent intervals required for the food handling spaces in the plant. Contamination or infestation in these areas is sure to be spread to the remainder of the plant.

The required numbers of toilets, washing facilities, urinals, and drinking fountains are usually specified in the local laws. One drinking fountain per 50 employees, one toilet and one hand washing fountain per 10 to 15 employees is a suggested ratio. The drinking fountains should be conveniently located, but not within toilet rooms or over hand-washing sinks. Walls of the toilets should be in light colours and should be washable. The rooms should be well ventilated with screened openings to the outside air. Hand washing facilities should be placed near the toilets. They are a visual reminder to wash hands after visiting toilets and before going to work. Additional hand washing facilities should be located near work areas. Frequent washing is an important dermatitis preventive measure. In a fruit processing plant, hand washing is essential because yeast and bacteria numbers and type in the product are of critical importance.

Individual towels (paper) or suitable drying devices (hot air), adequate soap (in dispensers) and covered waste receptacles should be available and serviced at regular intervals.

Dressing rooms should be designed for individual use and should be clean, well-lit and well-ventilated. If needed there should be 1 shower per 15 workers. Showers must be cleaned and treated with a sanitizing agent. They should be adjacent to the locker rooms and should be equipped with hot and cold water.

Health: Workers handling foods, or entering food handling areas should be required to have a thorough medical examination prior to beginning work, and this examination should be repeated at regular intervals. People who are afflicted with a communicable disease that can be transmitted through food, or are carriers of microorganisms that cause such a disease, should not work where food is being processed or served. Each employee has an obligation to report to his supervisor evidence of illness. Three groups of communicable diseases are common:

- Diseases which can be transmitted through intestinal contents (poor personal hygiene and poor hand washing after visiting toilets) such as:
  Amoebic dysentery (*Entamoeba histolytica*),
  Asiatic cholera (*Vibrio cholerae*),
  Bacillary dysentery (*Shigella dysenteriae*),
  Paratyphoid A (*Salmonella paratyphi*),
  Paratyphoid B (*Salmonella schottmuelleri*),
  Typhoid fever (*Eberthella typhosa*).
- Diseases which can be transmitted through mucous secretion of mouth and nose, such as:
  Staphylococcus aureus.
- Infection diseases such as jaundice, scarlet fever, diphtheria, and tuberculosis which can be transmitted through persons and food contaminated with such pathogenic agents.

Adequate supervision is necessary for detection of such conditions, and provisions should be made to remove such infected workers from fruit processing areas until a physician has certified that the danger of disease transmission is past.

Cleanliness and Hygienic Practices

Every human is a potential carrier of food-spoiling (and disease-producing) microorganisms in and on his body. For this reason, it is recommended that not only workers, but visitors in fruit processing areas must wear special protective outer clothing at all times when in these locations. The complete uniform includes a cover for the hair, and foot-wear (usually rubber boots or safety shoes). Freshly laundered uniforms will encourage employees to put on clean clothing to maintain a high level of personal cleanliness.

Nevertheless, the foremost requirement of personal hygiene is that the workers hands be washed thoroughly with soap and warm water and dried with hot air or with paper towels before starting work, after handling contaminated fruits, and after eating and especially after using the toilet. Drying the hand with a clean single-service towel or a stream of hot air is an obligation. Addition of disinfectants (such as chlor-hexidine or a quaternary ammonium compound) to hand soap or rinse water is not necessary. They may act as preservative for liquid soap, which could otherwise become a source of contamination. Frequent cleaning of empty dispensers and filling with fresh liquid is probably a more reliable means of controlling growth than the use of disinfectants.

The use of gloves of rubber or plastic is not recommended in tropical fruit processing plants. They are beneficial only when changed frequently and washed inside and out before being worn again. Otherwise, gloves themselves can become soiled so may only give an illusion of good hygiene. A feeling of awkwardness, the nuisance of changing, and the accumulation of perspiration on the hands are disadvantages that restrict their general use in tropical countries.

7.5 Pest Control

Several species of pests may occur in tropical fruit processing plants, but the most important are the following.

- *Flies:* House fly (*Musca domestica, M. sorbens*)
  Fruit fly (*Drosophila melanogaster, D. repleta*)
- *Beetles:* Dried fruit beetles (*Carpaphilus spp.*)
- *Ants:*Garden ant (*Lasius niger*)
  Pharaoh's ant (*Monomorium pharaonis*)
- *Cockroach:* American cockroach (*Pepriplaneta americana*)
  German cockroach (*Blatella germanica*)
  Oriental cockroach (Blatta orientalis)
- *Rodents:*Common rat (*Rattus norvegicus*) or brown rat
  Ship rat (*Rattus rattus*) or
  blackrat or roof rat
  House mouse (*Mus musculus*).

The aim of pest control should be to prevent infestation from developing "Preventation is better than cure". Insect and rodent proofing prevents physical damage to products, prevents added microbial contamination, prevents addition of filth to product, and reduces requirements for use of insecticides and rodenticides. Insects and rodents can also transmit diseases to humans.

Early detection and regular control procedures are necessary in plant and warehouses. Maintenance of pest control is of paramount importance.

The development of pest infestation in fruit processing plants depends mainly on the availability of food and harbourage with undisturbed surroundings. If the principles of good sanitation and good warehousing are put into practice, the conditions suitable for pests will not exist. Thus preventive measures are by far the most important aspect.

In order to minimize the presence of insects and rodents following aspects must be considered:

- Design and layout of plant and equipment,
- Insect- and rodent-proofing,
- Environmental sanitation,
- Good warehousing, and
- Maintenance of pest control.

Design and Layout of Plant and Equipment

Good design of building, processing lines, machinery, equipment, furniture, and storekeeping can do much to prevent infestation by insects and rodents from arising. On the other hand, it is far less expensive to build insect- and rodent-proofing into a building during construction than to do it after construction.

The aims should be to prevent or eliminate every space which might afford rodents harbourage, and every "dead" space and cavity which is difficult to clean and will provide harbourage for insects. Examples are double walls, floors, ceilings, ventilating equipment, underneath machinery and fixtures, the areas between machines and walls, insulated linings and below badly sited pipes.

Rounded wall/floor angles make cleaning easier and reduce the likelihood of dirt and residues accumulating in them. Structures must be built of materials impervious to rodent assaults. Outside doors should be made of metal or hardwood. If softwood doors are used, metal kick plates should be provided. Doors should not have any openings greater than 1 cm when closed. They should have double side and bottom guides to assure a permanently tight fit.

All outside openings from heating and ventilating equipment should be covered by rodent-proof grills.

Windows should be provided with tight-fitting screening with mesh openings of at least 16 to the inch to exclude houseflies and other insects. The complete screening of a fruit processing plant is not feasible unless the windows have been specially designed for it.

An ingenious method of preventing the entry of many flies is to provide a curtain of air moving across a doorway. Experiments show that an air speed of 7.62 m s^{-1} will exclude about 75 per cent of flies. Higher velocities were more effective, but objectionable to people passing through the barrier. The use of such air currents over openings is recommended for tropical fruit processing plants, especially jam factories. The use of screens on windows and doors is difficult and expensive. Weatherproof wire screens must be fitted outside all windows and this reduces light and ventilation and precludes the use of hanged windows which open outward. Two pairs of swing doors separated by a trap porch are required for external communication, to prevent entry of flies.

Light colored decorations (walls should be painted in light colors and should be washable) assist cleaning as dirt and residues are more obvious against a light background. Also, as insects normally prefer to breed in dark areas, a light background will discourage their development.

The surroundings of tropical fruit processing plants should be orderly and well-drained. Buildings should be preferably oriented so that they are well ventilated by the wind to avoid excess heating by the sun during the summer. In Egypt this would be east-west. All rooms should be adequately ventilated to help prevent condensation of moisture. Temperature and humidity play a part in pest infestations, particularly where certain insects species are concerned.

Building foundations should be brought to about 1 m above ground to prevent rodents from getting into the building and to reduce dust or other contaminants.

All wall junctions should be round, joints should be sealed to facilitate cleaning.

The processing, packaging and warehousing areas should be separated by partitions.

Light attracts many flying insects. If outdoor lighting is required, place lights away from the plant so they are not directly above windows or doorways. Inside the plant, lights should not hang directly above processing areas, as insects flying to these lights could fall into the product. Place light traps so they will draw flying insects away from processing areas. Clean out the light traps regularly to avoid the buildup of dead insects which serve as food for more serious pests.

Insect- and Rodent-Proofing

Pest-proofing consists of changing the structural details of a building to prevent entry of rodents and insects. It is not usually a practical proposition. However, in tropical fruit processing plants screens should be used to prevent flies and wasps from entering the plant. The maximum size of these screens should be 10–16 mesh (holes per linear inch) with 32 swg (Standard Wire Gauge = 0.3 mm or 0.012 in).

Rodent-proofing is more practical to prevent the access of rats and mice. Openings as small as 0.5 inch will admit young rats, while 0.25 inch openings will admit young mice. If only brown rats are present, rat proof all ground floor openings. If black rats are present, upper floors must also be treated. Use materials through which the rodents cannot gnaw. Gaps between doors and floors should not be more than 0.25 inch. The points at which service pipes, ducts and electric conduit conduits pass through the structure should be sealed. Broken glass or chicken wire mesh can be stuffed into holes which are then sealed with concrete.

A concrete floor is more effective for controlling rats than a curtain wall, and it increases the value of the building. Therefore, it always should be suggested in lieu of a curtain wall.

In the course of pest-control, may situations other than those dealt with will be encountered. The procedures outlined above are adaptable in most cases. Individuals directing proofing work should use their judgement regarding individual premises, remembering that nonessential work increases costs without increasing effectiveness.

Environmental Sanitation

Perhaps one of the more effective means of preventing insect and rodent infestation of a tropical fruit processing plant is to keep the premises, inside and out, clean and free of trash, garbage wastes, unused equipment, moist organic material, and filth which will attract pests to the site. Rapid and proper removal of wastes, garbage and trash prevents not only attraction of insects and rodents, but also prevents spoilage of wastes and garbage which may in turn contaminate process lines.

Cleaning of plant and equipment should be thorough, frequent, and regular. Regular attention should be paid to drains and gullies for they can be an important source of infestation by cockroaches, flies and rats.

Rubbish should be removed frequently and kept in closed containers which must be cleaned regularly. Dustbins are also important breeding sites for flies, and cockroaches if they are not maintained in a good hygienic condition. Poor stacking and storage arrangements and disorderly plant surroundings can also be attractive to rodents.

Good Warehousing

Storekeeping areas for raw materials and for finished products require the strictest insect and rodent control. Storage areas are an important source of infestation by cockroach, ants and rodents, and are also places, where they can even develop and breed. Therefore, all raw materials entering the plant should be inspected, for pests which may be carried in the goods. Signs to look for include rodent droppings, gnawed materials, damaged sacks, cartons, insect frass, insect webbing, live or dead insects, parts of "skins" of insects.

The key to a successful pest control operation in storage areas of finish products is early detection. Evidence of rodent infestation are: sighting of live or dead rats or mice, droppings, runways, tracks, gnawed material, nests, odour, rat urine (by UV-light) and broken containers. Evidence of insects such as ants and flies are more obvious. Visits during the earlier part of the night will usually reveal many cockroaches wandering about in search of food. In daytime, cockroaches will be hiding.

If evidence is found then appropriate action should be taken. This will depend on a number of factors, correct identification of the pest species being on of them. A knowledge of the basic life history and biology of pests is very valuable.

Segregation of goods is important, particularly raw material which should be stored quite separately from finished products. Generally the higher the temperature and humidity in the warehouse the more suitable are conditions for insects to breed and develop. Good ventilation helps to control temperature and humidity. Rat and mice also prefer warm conditions, but they are quite capable of breeding at low temperatures, e.g. in cold stores.

Stacks of goods should be built away from walls (at least 50 cm) to prevent accumulation of spoilage, to facilitate cleaning, use of insecticides and rodenticides, and inspection.

Systematic stock rotation should be practised so that older stocks are processed or sold first. "First in, first out" can be regarded as the guiding principle. Stores should, where possible, be completely emptied and thoroughly cleaned and disinfected prior to refilling. From the point of view of pest control this will help to prevent the build-up and spread of infestation from one lot of goods to others.

All floors, shelves, racks, cabinets and pallets require regular cleaning, especially if residues of fruits or raw material exist. Keep empty containers as clean as those storing food. Rejected products and damaged goods should be handled with the same care as the finished product.

Storage of products, utensils and equipments in toilet rooms, vestibules or under exposed sewer or water lines (except automatic fire protection sprinkler heads) is prohibited. Store pesticides, rodent poisons, cleaning agents, caustics, disinfections and other poisonous substances, in closed containers that are labeled in a way that readily identifies its contents. Each should be stored in a separate cabinet so that they are physically separated from each other.

Locker, toilets, offices, cafeterias or lunchrooms should also be routinely inspected for evidence of insects and rodents.

Maintenance of Pest Control

Chemical control methods are the commonest and most important methods for controlling flies, ants, cockroach and rodents. They include: insecticides (aerosols, lacquers, baits, sprays) rodenticides (acute rodenticides, chronic rodenticides) and residual treatments.

Since many infestations occur in places where food is prepared and stored, it is important to chose reasonably safe chemicals and use them with great care. It is a toxic and complicated procedure and can only be carried out by specialists. If the infestation problem is large, there are a number of Consulting-Bureaus who specialise in pest control. Large firms might well consider having a member of staff specially trained in sanitation and pest control.

It must be stressed that insecticides are not a substitute for cleanliness and good hygiene. Unless fruit residues, sugars, dirt, and rubbish are removed first, their application will not be effective, for the insecticidal deposit will remain on top of the dirt while insects will continue to live and breed underneath. Although there may be some control it will not be complete.

Physical methods of insect control include the use of heat, cold, radioactivity, centrifugal force and electrocution. The only one of these methods which has fairly wide application in fruit processing plants is electrocution. High voltage electricity (4500 V at 10 mA) is supplied to alternate wires of a grid, and insects settling on the latter are immediately incinerated. This involves the use of ultra violet radiation, sometimes called "black light" to attract flying insects to the high voltage grid. A tray is placed below it to collect the electrocuted insects, thus preventing them from falling into food or machinery.

Break-back traps can be used for rodent control where the infestation is very small. Their biggest advantage is that they involve no poison risks. On the other hand, mice and rat eradication with ultrasound is now widely used. When put under stress by the continuous application of the unbearable sound, the rodents will leave the rooms. The range of effectiveness is from 150 m^2 to 20 000 m^2 depending on the unit used.

For those who want to know more about insect and rodent life history, biology, identification and control several references are available[115–123].

8 Waste Disposal Control

Waste disposal is one of the items to be determined before the fruit processing plant is built. It is not only inevitable, but now there are many laws and regulations which must be compiled with. Failure of a fruit processor to plan accordingly could eventually result in excessive costs for treatment or disposal of his wastewater. Waste disposal may even determine at what location the plant will be built.

Waste disposal control can be a special task for the quality assurance manager or be a department for itself having a relation to the quality assurance department.

This chapter will indicate the data that a processor could collect which would enable an engineer to construct a proper disposal system and to solve the problems which arise. Some of the following suggestions may be useful in ideal circumstances, but sometimes may be impractical for reasons of economy. If there are any serious difficulties the problem should be referred to a competent consulting bureau.

This chapter describes briefly the general wastewater characteristics associated with the production of fruit products, methods of waste reduction and waste utilization. It includes the application of current technologies now available for treatment of fruit wastes, and their advantages and disadvantages. Methods of the analysis of wastewater are presented in detail.

8.1 Definition and Terminology [124–137]

This section indicates some terms usually used in waste disposal control, wastewater control and treatment:

– *BOD*, Biological Oxygen Demand, is a measure of the concentration of biologically degradable material present in the waste. It is the amount of free oxygen utilized by aerobic organisms when allowed to attack the organic matter in an aerobically maintained environment at 20 C for 5 days. It is expressed in milligrams of oxygen utilized per liter of liquid waste volume (mg/l) or in milligrams of oxygen per kilogram of solids present (mg/kg = ppm).

– Biological treatment, organic waste treatment in which bacterical (aerobic and anaerobic) action is intensified under controlled condition.

– Chemical treatment, is chemical precipitation using chemical coagulants of lime followed by ferrous sulphate or alum in a batch or continuous flow system.

– COD, Chemical Oxygen Demand, is determined by the amount of potassium dichromate consumed in a boiling mixture of chromic and sulphuric acids. It is an indirect measure of the biochemical load. The amount of oxidizable matter is proportional to the potassium dichromate consumed. Here the waste contains only readily available organic bacterial food and no toxic matter (as in the waste of fruit processing plants), the COD values can be correlated with BOD values obtained from the same wastes.

– Composting, is the aerobic, thermophilic decomposition of organic wastes to a relatively stable humus. The decomposition is done by aerobic organisms, primarily bacteria, actinomycetes and fungi. The resulting humus may contain up to 25% dead and living organisms.

– Contamination, also pollution, is the presence of microorganisms or substances in water (or soil or air or in the product) in such quantities that the natural quality of water etc is degraded.

– Filtration, is the process of passing wastewater through a porous medium for the removal of suspended or colloidal material by a physical straining action.

– Imhoff tank, is a deep two-storied sewage tank patented by Karl Imhoff consisting of an upper sedimentation chamber and a lower sludge digestion chamber. The floor of the upper chamber slopes to trapped slots, through which solids may slide into the lower chamber; the digestion process is anaerobic.

– Lagoon, is a term commonly given to water impoundment in which waste water is stored or stabilized or both. Lagoons may be described by the predominant biological characteristics (aerobic, anaerobic or facultative), by location (indoor, outdoor), by position in a series (primary, secondary, or other) and by the organic material accepted (sludge, sewage or other).

– Screening, is a preliminary treatment for removal of large-particle solids prior to final treatment, discharge into receiving waters, or discharge into a municiple sewage system.

– Sedimentation tank, is a tank or basin in which wastewater containing settleable suspended solids is retained for a sufficient time so that part of the suspended solids settle out by gravity. The time interval that the liquid is retained in the tank is called *detention period*. In sewage treatment, the detention period is short enough to avoid *putrefaction*. Putrefaction is the decomposition of protein and the production of offensive odours.

– *Septic tank*, is a single-story tank in which the organic portion of the settled sludge is allowed to decompose anaerobically without removal or separation from the bulk of the carrier water flowing through the tank. Only partial liquefaction and gasification of the organic matter is accomplished, and eventually the undecomposed solids will accumulate to the extent that removal of the solids is necessary.

- *Settleable solids*, are the suspended solids that will separate when the waste water is held in a quiescent condition (such as an Imhoff cone or cylinder) for 1 hour. It can be recorded as milliliters/liter/hour.
- *Sewage*, is water after is has been fouled by various uses. From the standpoint of source it may be a combination of the liquid or water-carried wastes from residences, business buildings, and institutions, together with those from industrial and agricultural establishments, and with such ground water, surface water, and storm water as may be present.
- *Sewerage system*, is a system for the collection and disposal of sewage or industrial wastes of a liquid nature, or both, including various devices for the treatment of such sewage or industrial wastes.
- *Sludge*, is the accumulated settled solids deposited from sewage or other wastes, raw or treated, in tanks or basins, and containing more or less water to form a semi-liquid mass.
- *Spray irrigation*, or land disposal, is an economical and unobjectionable waste disposal method when land is available. A vegetative cover crop is essential in order to increase the rate of absorption and transpiration and to prevent soil erosion.
- *Trickling filter*, is a process used in wastewater treatment as a method of contacting dissolved and colloidal organic matter with biological active aerobic slime growths. It is not a true filtration process.

8.2 Factors To Be Considered in Waste Disposal Control [124–137]:

The quality assurance department should investigate the following basic considerations in the interests of simplicity of design and economy of operation.

8.2.1 Character of the Waste

It should be determined if the waste contains materials hazardous or potentially hazardous to public health (e.g. pesticides, heavy metals ...). It is also desirable to know the organic matter content (determined as biochemical oxygen demand BOD and chemical oxygen demand COD), temperature, pH, total suspended solids, settleable matter, and amount of suspended solids which could be removed by screening.

In general, fruit processing waste does not transmit disease and does not contain materials hazardous to public health. The high organic strength of normal fruit wastes (BOD = 2000–3000 mg/l: COD = 4000–8000 mg/l; settleable matter = 5–30 mg/l and pH = 5–6) is the principle reason for the difficulty encountered in their disposal. As compared to domestic sewage, fruit processing wastes are high in sugars and the biochemical oxygen demand is about ten times greater.

Raw untreated wastes consist of small fruit particles and sometimes discarded whole pieces of raw products, peels and seeds; and also a high percentage of soluble organic materials. As long as oxygen is present, decomposition of the organic material will proceed without occurence of the stream conditions usually associated with pollution. However, if the strength and volume of the waste water is such that dissolved oxygen disappears from the water, then fish and other forms of aquatic life will disappear. Foul odours will arise from sludge deposits and floating scum.

Another problem with waste from a fruit processing plant is the large seasonal waste and that solid waste is not possible to store. Seasonal production and wide variation of fruit means that wastewater treatment can be costly because of equipment lying idle during the rest of the year and variations in effluent strenght complicate the problem. Fruit washing, peeling and blanching are the main sources of solid waste. On the other hand, much of the total waste water comes from relatively clean sources such as container washing, cooling water and evaporator condensing water, which can be reused.

As the fruit wastewater is generally characterized by its containing a high percentage of soluble organic matter and is biodegradable, the biological treatment of waste water is considered the best practical method. Generally the wastes are amenable to treatment together with domestic sewage at municipal sewage treatment facilities.

8.2.2 Waste Flow

Because the quantity and character of waste discharged from fruit processing plants are highly variable, there can be no substitute for a plant survey to determine the actual waste characteristics and loads. Perhaps the single most important wastewater characteristic to dertermine during the waste survey is flow. This because flow is the limiting design factor used to calculate the size of many treatment units. On the other hand, the gathering of accurate flow and waste characterization data is the most important step the quality assurance manager can take to ensure that waste treatment facilities are not too small.

An automatic metering and recording of wastewater flow is recommended. The data should be calculated in terms of daily, weekly and yearly volume of waste, such as:

Wastewater flow
    hours/day
    days/week
    weeks/year

Table 34. Character of Waste Water [124]

		Maximum value	Average value	Minimum value
1. Waste water flow	hours/day days/week weeks/year	*)		
2. Waste water quantity	m^3/hour m^3/day m^3/week	*)		
3. COD-concentration	mg/l			
COD-concentration settled	mg/l			
COD-freight	kg/hour kg/day kg/week			
4. BOD-concentration	mg/l			
BOD-concentration settled	mg/l			
5. Some important anorganic components:				
– ammonium	mg/l			
– calcium	mg/l			
– sulphate	mg/l			
– NaCl	mg/l			
6. Settleable solids	ml/l			
Floatable solids	mg/l			
7. pH-value				
8. Temperature	C			
9. Anorganic components:				
– sulphur – total value	mg/l			
– nitrogen – total value	mg/l			
– phosphorous – total value	mg/l			
– nitrate	mg/l			
– magnesium	mg/l			
– heavy metals (Fe,Ni,Cu...)	mg/l			

*) periods; yearly and/or daily progress lines

Wastewater quantity
   m^3/hour
   m^3/day
   m^3/week
COD – freight
   kg/hour
   kg/day
   kg/week

Tests of pollutant loads commonly found on the basis of raw fruit used are:

Average flow
   0.15 m^3/tonne
   0.8–2.0 kg BOD_5/tonne
   1.5–3.8 kg COD/tonne

8.2.3 Segregation of Highly Contaminated Wastewater

The possibility should be considered of separating individual waste flows at the point of origin into contaminated water requiring treatment, and waters with little or no contamination which could be discharged without treatment. The latter group would include cooling waters, condenser water, etc. This is an important point in the initial survey of the problem because the volume of waste waters requiring treatment determines the size of the treatment plant. Any reduction in the volume of the waste is an ultimate saving in costs.

8.2.4 Reduction of Waste Flow

A program of waste preventation, waste reduction and reusing of water not only reduces the amount of waste and waste water, but results in a saving through greater production utilization whithin the plant, in reducing the total production expense and saving in total water treatment costs. The reduction in volume of the waste water may not be accompanied by a reduction in the total organic strength. It usually causes changes in waste quality and character and also concentration of the waste loads in certain process streams. However, it will make treatment of the waste more economical and effective. Water reuse, however, ought not to be indiscriminate. Full regard should be given to the fact that improper reuse of water may lead to bacteriological problems.

Following suggestion for waste prevention, waste reduction and water reuse are important:

– Wash water can be continuously recycled within an optimum system.
– Cooling tower recirculation is a common method of conserving water.
– The cooling tower blowdown is used to supply water to the various subprocesses of the processing line.
– Reducing the fresh water requirements, by means of recycling and controlled conditions in fruit washing and container washing equipments.
– Using the counter flow principle, in which water from the last product washing, instead of being wasted, is collected and fed back in a counter flow system to be used in the preceding washing operation. Thus, the product is moved forward after each washing operation, into water which is cleaner than that used in the preceding treatment.
– Keeping all sanitary fittings, sanitary valves, and filler valves in good repair.
– Repairing or replacing leaky containers.
– Reducing waste amounts as far as possible by modification of in-plant processes.
– Dry handling and disposal of solid waste from machines and floors.
– Limiting unnecessary use of water.

- Choosing improved systems for removing suspended solids from waste water. The liquid and solid wastes should be handled separetly.
- High pollutional loads are generated by conventional blanching (Thermobreak) in tropical fruit processsing plants, using hot water. With this method nutrients and minerals are also leached and destroyed. Using steam or possible microwaves, can reduce such problems.
- Maximum by-product recovery and waste utilization as far as possible,

8.2.5 Solid Waste Disposal and Utilization

The disposal of wet solids from a tropical fruit processing plant is a major problem. The solid waste is similar to waste generated in household kitchens, it can cause problems if it is not handled promptly and properly. Prompt disposal is necessary to minimize the propagation of insects and rodents, and to eliminate odours associated with the decomposition of organic matter.

Complete utilization is the ideal solution to the solids disposal problem. This increases profits and reduces air and water pollution. In some cases fruit waste solids have been successfully turned into dehydrated stock feed. Commercial pectin production from mango peels is possible (jelly grade 200–240). Mango kernel fat obtained by hexane extraction finds use in soap making and cocoa-butter substitution. Kernel powder can be also used as cattle feed. The enzyme bromelin can be extracted from pineapple waste. The press cake after drying may be used as cattle feed. It is also technically feasible to produce alcohol from fruit waste or to use it for yeast production.

The major wastes from tropical fruit processing plants are not utilized in an organized manner in comparison to apple or orange processing plants. The solid waste from orange which is produced in large quantities over a large season is now used for number of by-products, such as: pectin, essential oil, pulp wash, stock feed, ...

The possible reasons for non-utilization of tropical fruit wastes are:

- Tropical fruit processing plants are usually small scale units whose scale of operation is such that the amount of waste available is negligible.
- The fruit processing is highly seasonal and the season being very short, the processors feel that they will be working throughout the day and they cannot spare time to attend to the processing of waste.
- The solid waste have relatively little economic value.
- The cost of the plant and machinery for processing waste and technical know-how is often much more than the primary processing facilities.

Generally, most tropical fruit processing plants discharge their solid waste by spreading over farm land, or by landfill. However, it must be under careful sanitary control to prevent fly breeding, odour problems and possible ground water contamination. Composting is a widely used method of disposing of solids such as peels. It is a method of converting plant residues to more stable organic materials, which are valuable as feed, as fertilizers or soil conditioners. Also, disposal may be by removing to dry beds from which it is later piled and burned. In some urban areas processors have to pay on a tonnage basis to have the solids removed. In these instances it is desirable to have the waste as dry as possible. Vibrating type screens give drier solids than revolving drum screens. Elevation of the solids to a hopper facilities loading into trucks and allows considerable dewatering during storage in the hopper.

Daily washing and cleaning of putrescible-waste containers is a necessity. A high pressure hydrant hose, detergents and disinfectant should be provided. In some instances insecticides are needed to prevent flies, bees and other insects from interfering. The containers must be watertight, corrosion resistant and easily cleaned.

8.2.6 Existing or Proposed Regulations Governing Waste Disposal

Food plant wastes normally have to be discharged to an inland river or to the sewer of a local authority. These discharges have to conform to standards of quality imposed by regulating authorities. The standards vary considerably from country to country and may be appreciably different in different parts of the same country. National regulations are generally less stringent than local or province restrictions. This is because the specific requirement to protect the water quality of a receiving stream or sea or to ensure the protection of a city's wastewater treatment system is often more limiting than the broad-based national standards. For effluents discharged to inland rivers the quality required is likely to be highest where the dilution afforded by the stream is small, or where the water supports a fishery or is used for domestic supply. Generally, wastewater from tropical fruit processing plants do not contain substances toxic to bacteria or to fish. Their constituents are mainly organic substances which are nonetheless objectionable in public water since organic matter is a source of food for microogranisms which reduce the dissolved oxygen in the water when breaking down the organic matter.

The general standards usually have the following limits:

- Temperature: < 35 C.
- pH-Value: 5–9
- Settleable substances: < 10 ml/l after 0.5 hours.
- Suspended solids: < 20–40 mg/l.
- COD-value: < 100 mg/l.
- BOD_5-value: < 20–30 mg/l.

In the mid-1970s national standards were established in the USA for canned fruit processors. These standards required that by July 1977 the industry was to achieve a reduction of pollutants equivalent to secondary treatment or levels referred to as *"Best Practical Technology Currently Available"* (BPT). The same national regulations established a target date of 1983 to achieve further reductions in pollutants, referred to as *"Best Available Technology Economically Achievable"* (BAT). Limits for BPT and BAT were established for both the peak day and 30-day average allowable discharge loads. A partial list of both BPT and BAT limitations for some canned fruits is given below [125]:

Table 35. BPT and BAT values

| Category | BPT (kg tonne^{-1}) | | | | BAT (kg tonne^{-1}) | | | |
| | BOD | | TSS | | BOD | | TSS | |
	Peak day	30-day average	Peak day	30-day average	Peak day	30-day average	Peak day	30-day average
Apple juice	0.60	0.30	0.80	0.40	0.20	0.10	0.20	0.10
Apple products	1.10	0.55	1.40	0.70	0.20	0.10	0.20	0.10
Apricots	3.00	1.81	5.36	3.74	1.26	0.94	1.26	0.94
Citrus products	0.80	0.40	1.70	0.85	0.14	0.07	0.20	0.10
Dried fruit	1.86	1.13	3.34	2.34	0.73	0.56	0.73	0.56
Grape juice	1.10	0.60	1.99	1.44	0.77	0.58	0.77	0.58
Peaches	1.51	0.93	2.72	1.03	0.77	0.58	0.77	0.58
Pears	1.77	1.12	3.21	2.32	0.86	0.66	0.86	0.66
Pineapples	2.13	1.33	3.86	2.76	1.48	1.11	1.48	1.11
Plums	0.69	0.42	1.24	0.87	0.28	0.20	0.27	0.20
Tomatoes	1.21	0.71	2.16	1.48	0.52	0.38	0.52	0.38

TSS = Total Suspended Solids.

At present the US government is in the process of reevaluating the BAT limitations for the fruit industry. The proposed regulations (possibly less stringent that the recently withdrawn BAT levels) are referred to as *"Best Conventional Pollutant Control Technology"* (BCT).

In addition to national limits based on kilogram per day, more stringent standards required by local agencies usually include limits on the concentration of pollutants.

8.2.7 Selection of Treatment and Disposal Procedures

The main reason for waste treatment is to comply with wastewater discharge standards. The high capital costs of wastewater-treatment equipment have been a major factor in locating fruit processing plant in areas where discharge to a river or to city sewers is feasible. Most of the fruit processing plants discharge wastes to city sewers. Some plants discharge directly to land disposal systems, and only a few have their own wastewater treatment systems that discharge directly to receiving water.

The decision to own and operate a wastewater treatment system can be costly. However, the cost of discharge to city sewers means that many processors must install some degree of treatment. The exact extent of that treatment varies from screening (the cheapest method) to more costly methods.

At present there are two basic ways of treating fruit processing wastes prior to discharge into streams. They are called primary and secondary. *Primary treatment* is designed to reduce or to remove settleable wastes and some suspended solids. The methods used are screening, centrifugation, flocculation or sedimentation or a combination of these. The BOD in the form of suspended solids may be reduced by 70%.

Secondary treatment is a more refined treatment and may include screening and settling, but uses biological processes to reduce dispersed solids and soluble organic content of liquid waste. This treatment included activate sludge, trickling filters or extended aeration. It can remove from 70–90% of the suspended solids and reduce the BOD by 80–90%. There are other, more advanced methods of treating wastes which are generally called tertiary or special treatment methods. These include treatment of the effluent in oxidation lagoons, filtration through activated carbon or synthetic exchange media, membrane separation or an ion exchange system for salt removal. Such treatments are seldom used on account of high cost.

The following is a synopsis of present treatment methods common throughout the fruit processing industry.

Screening

Screening is employed as preliminary treatment for removal of large particle solids prior to final treatment, discharge into receiving water, or discharge into a municipal sewage system. Screen sizes of 20 to 40 mesh are commonly used. Three types of screens are typically used: static, rotating or vibrating.

In *static screens* wastewater from raisin manufacture passes through a tangential or static screen while coarse solids and denser material slide tangentially into a solid hopper. A advantage of static screens is that no moving parts are required, thus maintenance is minimal. However, care must be taken in screen selection, otherwise the raw wastewater may also shoot tangentially into the solids hopper.

Rotating screens are essentially revolving drums covered with a screen cloth. They are especially useful when a high solids capture rate is required or fine mesh screens are used. The wastewater is discharged to the outside of a rotating screen drum. The solids are scraped from the drum by a doctor blade, while the screened effluent passes through the drum and is discharged from the drum's centre. Advantages of the rotating screen are that it achieves a high rate of solids removal and is self-cleaning. The solids are usually removed to a hopper by screw conveyor or bucket elevator. The rotary drum screens of Passavant-Werke AG, D-6209 Aarbergen 7, which is called Rotopass, can be recommended.

Vibrating table type screens are now widely used when a large volume or high rate of solids must be screened. Finer screen cloths can be used than would be possible with a drum type screen, and the amplitude of the vibrations can be adjusted for differences in the material being screened. Vibrating screens may be purchased in different sizes. The size and number of screen units needed will depend on the volume of waste to be screened.

Sedimentation

Sedimentation and settling in collecting ponds or lagoons is usually employed to remove suspended or settleable solids from tropical fruit processing waste effluent prior to the discharge of the waste for further treatment. The basins are often equipped with overflow weirs and baffles for continuous operation and with means for addition of chemical agents to facilitate flocculation of the particulate matter.

Ordinarily chemical treatment for fruit processing waste would be undertaken only after careful consideration of the cost, the inconvenience of handling the large volumes of sludge produced, and the failure of the method to give a reduction in strength of the wastes of more than about 50%. Controlled chemical treatment will remove suspended and colloidal solids but will not affect solids in solution (such as sugar). Chemical precipitation, at best, removed from 25 to 50% of the BOD. In wastewater having a very high level of total dissolved solids, the reduction in BOD in the effluent is insufficient.

Two types of chemical treatment are used: the continuous flow method, and the fill-and draw or batch type treatment. Although the continuous flow type of treatment will handle larger volumes of water in a given period of time, it is difficult to maintain the optimum chemical dosage, and it is difficult to remove the large volumes of sludge produced. The fill-and draw method largely overcomes these disadvantages. With this method screened waters are pumped into one of two tanks. When this is full the flow is turned into the second tank. Agitation is started in the first tank and the proper amount of the first chemical (lime 4 kg/1000 l) is added. Then half the dosage of the second chemical (aluminium sulphate) 1 kg/1000 l or Ferrous sulphate (4 kg/1000 l) is added while agitation is continued. After several minutes the remainder of the second chemical is added slowly until a large, heavy flocs begin to settle out. When the supernatant liquid is clear it is discharged and the sludge at the bottom is pumped onto sludge drying beds. Ferrous sulphate reacts with hydroxyl ions to form hydrous oxides. The latter are relatively insoluble at normal pH values and tend to flocculate, coalesce and settle. Zinc chloride (2 to 6 kg/ 1000 l) can be also used in second treatment.

The biological material can be oxidized from the wastewater at the rates of 1.8 parts per part of potassium permanganate per hour, and 1 to 6 parts per part of chlorine (from NaOCl) per hour, these treatment were considered simple but expensive.

Centrifugation

The simplest type of wastewater purification is mechanical separation of the solids by means of screens and decanters. The wastewater is first screened, so that coarser solids, such as parts of pulp and peels, are continuously removed. The preclarified wastewater flows into the decanter. The decanter concentrates the solids still contained in the waste water and discharge them continuously with a high dry substance. Dewatering of the solids take place simultaneously with separation.

Using decanters for separating highly dispersed and suspended material is possibly more effective than screening but more expensive. A large percentage of settleable solids are removed from the effluent and a COD reduction of 50 to 70% should be accomplished if the effluent has been properly screened. The decanter is a horizontally arranged centrifuge with a cylindroconical solid-shell bowl which is used for the continuous removal of solids from suspensions. Through the centrally arranged feed tube, the wastewater enters the separation chamber of the bowl where it is brought up to operating speed. The solids, being subjected to the high centrifugal acceleration, settle against the bowl wall almost instantly. The conveyor screw, rotating at a slightly higher speed that the bowl shell, continously delivers the centrifugally removed solids to the small diameter end of the bowl. Due to the conical shape of the bowl, the solids are "lifted" out of the liquid and, while passing through the "drying zone", the liquid is removed by centrifugal force. The solids are then discharged through ports at the end of the bowl into the catcher of the housing. The liquid flows between the screw spirals towards the other end of the bowl. While passing through the "clarifying zone", specifically lighter particles still present in the liquid are centrifugally removed and conveyed by the screw to the solids outlet together with the solids removed in the inlet zone.

Decanters from the following companies are available:

– Westfalia Separator AG
– Alfa-Laval AB
– Flottweg Werk, Bird Machine GmbH
– Sharples-Stokes Division.

Flocculation will increase the removal efficiency of clarifiers. Mechanical flocculation is accomplished by gently stirring the liquid with rotating paddles, causing the finely divided particles to coalesce, with improved settling properties. In some cases it is necessary to add chemical flocculation agents such as lime, aluminium sulphate, copper or various ferric salts, following by gently stirring.

Disposal of Waste by Spray Irrigation

The most highly recommended method for liquid waste disposal of a tropical fruit processing plant is spray irrigation. In recent years spray irrigation has been increasingly used as a means of cannery waste disposal. It serves as an economical and unobjectionable waste disposal method when suitable land is available, especially when the processing plant is in a rural area where land is cheap, where no city sewer is available for pretreatment and municipal discharge of treated effluent. In many respects it is an improvement over other methods. It consists of spreading the screened wastewater over the land by means of a high-pressure sprinkler system.

Although land treatment has advantages, the quality assurance manager should realize that treatment by this method can appear deceptively simple. He must pay attention to contaminitation of groundwater, surface runoff, surface erosion, damage to the vegetation, loss of soil filtration capacity, and other environmental damage.

The following data are required in order to set up a successful spray irrigation system:

- Quantity of wastewater per day, week and season.
- Character of wastewater after the primary treatment. Beside the data on pH, dissolved oxygen, total suspended matter, settleable matter, BOD and COD, information about sodium content, to avoid poisoning of plants, and other minerals are important.
- Land available for disposal area. Its location must be within practical and economical pumping distance of the plant.
- Topography of the land and characteristics of soil. Spray irrigation of land which is not fairly level may not be successful because of run-off and erosion. Successful spray irrigation depends upon the capacity of soil to absorb the water, depth of ground water, infiltration capacity, content of organic matter, cation exchange capacity and the nature of the cover crop.
- Selection of cover crop. A cover crop is essential in order to increase the rate of absorption and transpiration and to prevent soil erosion. Cover crops include tall fescue, reed canary grass, orchard grass, mammoth clover and planted vegetable crops.

With a good cover crop and average soil absorption capacity, fruit processing waste can be successfully disposed of. The quality assurance manager should consult local agronomists, soil specialists and farm advisory staff for expert assistance. The farming operation may be concentrated to a private farmer.

The spray irrigation system usually requires the following items: a screening unit, collecting tanks, a pump, a main line for transporting waste to the irrigation site, lateral line for distribution from the main line, and self-activated revolving sprinklers.

The use of overland flow instead of spray irrigation is possible. Its application has the advantages of reduced land area requirements compared with spray irrigation method, reduced wastewater sprinkler or application equipment costs. Its disadvantages are the greater possibility of nuisance problems with increased waste loading, the land surface must have a uniform slope, generally from 2 to 8%, so that runoff will move in the desired direction, a series of mounds parallel to the ground slope to discourage the applied water from short-circuiting the full width of the field, and the need to discharge treated runoff.

Lagooning-Biological Disposal

Discharge of screened raw fruit processing wastewater into a municipal sewage treatment plant is the most desirable disposal method if satisfactory arrangements can be made. Spray irrigation is also recommended. However, the seasonal nature of fruit processing and the high pollutional strength of the wastes may cause serious problems.

Biological methods are most suitable for testing the part of wastewater with highest pollutional strength. For example waste water from the blanching operation. All types of fruit waste will support the growth of bacteria and the fermentation occurs very quickly and the end-products of the fermentation are compounds of much lower pollutional strength.

Both anaerobic and aerobic digestion of fruit-processing waste have been used in large scale operations. Although anaerobic treatment requires more elaborate equipment and more careful attention, it offers specific advantages over aerobic treatment:

– it uses less electrical energy (since oxygen is not required).
– it produces an usable byproduct, methane.
– there is less biological sludge to dispose of, which is today very important, and
– nutrient requirements are lower than with aerobic treatment.

On the other hand, wastewater treatment using aerated lagoons, modified activated sludge and trickling filters are recommended.

For selection the proper wastewater treatment the quality assurance manager should consult international companies working on wastewater treatment or the problem should be referred to a competent consulting bureau, e.g.:

– Aqua Consult Ingenieur GmbH
– Passavant-Werke AG
– Sulzer Chemtech.

For those who want to know more about waste disposal, waste reduction, waste management and waste treatment several references are available [97,126–137].

8.3 Methods of Analysis

Tests generally made on liquid waste of fruit processing plants are:

- temperature
- pH and acidity,
- dissolved oxygen,.
- total suspended matter,
- settleable matter
- biochemical oxygen demand (BOD),
- chemical oxygen demand (COD).

Temperature, pH, conductivity and dissolved oxygen can be measured using portable instruments. On-Line monitoring systems fitted with recorders are also available. Some companies producing such equipment are:

- Winopal Forschungsbedarf
- Fisher Scientific
- Obrisphere GmbH

More important for the examination of waste water is the determination of BOD, COD, and may be also the total suspended matter and settleable matter.

The methods outlined in this section have been adapted from "Standard Methods for the Examination of Water and Waste Water", American Health Association, 16th Edition 1985; except as otherwise indicated and referenced.

8.3.1 Total Suspended Matter

The definition of solids in water analysis is the remaining residue upon evaporation and drying at 103–105 C. Total suspended matter is the total nonfiltrable residue dried at 103–105 C. The total nonfiltrable residue is the retained material on a standard glass fiber filter disk after filtration of a well mixed sample of waste water. The residue is dried in an oven at 103–105 C for at least 1 h, cooled in a desiccator and weighed. Repeat the drying cycle until a constant weight is attained or until weight loss is less than 0.5 mg.

Generally, 100 ml (or a larger volume if total suspended matter is low) containing not more than 200 mg total nonfiltrable residue is filtered under vacuum over a glass fiber filter disk placed on the bottom of a suitable Gooch crucible. To prepare the filter disk, wash it with distilled water over the Gooch crucible, dry the filter combination at 103–105 C for 1 h, store in desiccator until needed, and weigh immediately before use.

Calculation:

$$\text{mg/l Total Suspended Matter} = \frac{(a-b)\,1000}{c}$$

Where:

a = weight of filter + residue
b = weight of filter
c = ml of sample

Suspended solids determination is one of the major parameters used to evaluate the strength of waste water and determination of the efficiency of treatment units. Suspended solids correlate positively with the BOD determination.

8.3.2 Settleable Matter

The determination of settleable solids is of particular importance since it offers a basis for prediction of sludge load in settling tanks, stream beds, sewer lines, or lagoons.

Settleable matter may be determined and reported on either a volume (ml/l) or a weight (mg/l) basis.

The test is ordinarily conducted in an Imhoff cone, although a graduated cylinder could be used, allowing 1 h settling time. Fill the cone or the cylinder to the litre mark with thoroughly mixed sample. Settle for 45 min, gently stir the sides of the cone with a rod or by spinning, settle 15 min longer, and record the volume of settleable matter as millilitres per litre. The practical lower limits is about 1 ml/h. Where a separation of settleable and floating materials occurs, do not estimate the floating material.

To determine the settleable matter by weight, determine firstly the suspended matter (mg/l) in the sample as described before. Pour a well-mixed sample into a glass vessel not less than 9 cm in diameter. Use a sample of not less than 1 l and sufficient to give a depth of 20 cm. A glass vessel of greater diameter and a larger volume of sample also may be used. Let it stand quiescent for 1 h and without disturbing the settled or floating material. Siphon 250 ml from the center of the container at a point halfway between the surface of the settled sludge and the liquid surface. Determine the suspended matter (mg/l) in all or in a portion of this supernatant liquor as directed before. This is the nonsettling matter.

Calculation:

mg/l settleable matter = mg/l suspended matter – mg/l nonsettleable matter.

8.3.3 Dissolved Oxygen (DO)

Dissolved oxygen (DO) levels in natural and waste waters depend on the physical, chemical, and biochemical activities in the water. The analysis of DO is a key test in water pollution and waste treatment process control. In carrying out the procedure for the BOD test, it is necessary to know the concentration of dissolved oxygen in the samples after 15 min. and after 5 days, incubation at

20 C. The sodium azide modification of the Winkler method is recommended for fruit processing waste-BOD determinations.

This test can be used to standarize an electronic oxygen monitor. This type of instrumentation, which is available commercially, is highly recommended because it eliminates the problems caused by suspended solids and water colour. Because the DO content is affected by temperature, barometric pressure, and dissolved solids concentration, it is necessary to make DO determinations of samples at the time of collection.

In the determination of DO, various materials cause interference. These include iron salts, organic matter, excessive suspended matter, sulphide, sulphur dioxide and residual chlorine.

Chemistry of the Azide Modification of the Winkler Method:

In the Winkler method, if no oxygen is present, a white precipitate is formed when manganese sulphate ($MnSO_4$) and $NaOH + KI$ are added:

$$Mn^{++} + 2\,OH^- \longrightarrow Mn(OH)_2 \text{ white precipitate}$$

When oxygen is present the Mn^{++} is oxidized to a higher valence and precipitates as a brown floc:

$$Mn^{++} + 2\,OH^- + O \longrightarrow MnO_2 + H_2O$$
$$Mn(OH)_2 + O \longrightarrow MnO_2 + H_2O$$

Shaking permits the oxygen to react causing the oxygen fixation (Mn^{++} to MnO_2). The additon of H_2SO_4 lowers the pH and MnO_2 oxidizes I in the alkali-iodide-azide reagent to free I_2.

The azide modification removes the interference caused by nitrate ions, often found in effluents of biological processes, which oxidizes I to free I_2 under acid conditions but does not oxidize Mn^{++}. It is particularly troublesome because its reduced from N_2O_2 is oxidized by oxygen which enters the sample during the titration and converts it to nitrite, NO'_2., establishing a cyclic reaction that will lead to high results. Sodium azide is added in order to complex the nitrite out of the reaction.

$$NaN_3 + H^+ \longrightarrow HN_3 + Na^+$$
$$HN_3 + NO^-_2 + H \longrightarrow N_2 + N_2O + H_2O$$

Reagents:

– *Manganese Sulphate Solution*, dissolve 480 g $MnSo_4 . H_2O$ in distilled water, filter, and dilute to 1 liter.
– *Alkali-iodide-azide reagent,* dissolve 10 g NaN_3 in 500 ml distilled water. Add 480 g sodium hydroxide ($NaOH$) and 750 g sodium iodide (NaI) and stir until dissolved.
– *Sulphuric Acid 36 N*, 1 ml N H_2SO_4 is equivalent to 3 ml of alkali-iodide-azide reagent.

- *Starch Solution*, make a thin paste of 20 g of soluble starch and 2 g salicyclic acid (for preservation) in a small quantity of distilled water. Pour this paste into 1 l of boiling, distilled water. Allow this mixture to boil for a few minutes, then cool and settle overnight. Remove the clear supernatant and save, discard the remainder.
- *Sodium Thiosulphate Solution 0.025 N*, dissolve exactly 6.205 g sodium thio-sulphate crystals (Na$_2$S$_2$O$_3$ x 5 H$_2$O) and 0.4 g of solid sodium hydroxide (NaOH) in freshly boiled and cooled distilled water and make up to 1 l. Prepare fresh weekly. Standard sodium thiosulphate solution exactly 0.0250 N (which is available commercially), is equivalent to 0.200 mg DO per 1.00 m.).

Determination of DO

To a 300 ml sample add 2 ml MnSO$_4$ solution followed by 2 ml alkali-iodide-azide reagent, below the surface; stopper to exclude air and mix by inverting. When the precipitate settles leaving a clear supernatant above the manganese hydroxide floc, shake again. When settling has produced at least 100 ml clear supernatant, carefully remove the stopper and add 2.0 ml concentrated H$_2$SO$_4$ by allowing it to run down the bottleneck, restopper, mix by gently inverting until dissolution of the floc is completed. The iodine should be uniformly distributed before decanting the amount needed for titration. This should correspond to 200 ml of the original sample after correction for the loss of sample by displace-ment with reagents has been made.

$$200 \times \frac{300}{300-6} = 204 \text{ ml}$$

It is recommended that you use a glass-stoppered BOD bottles with collar, 300 ml, and DO–BOD volumetric flask graduated to deliver 201 ml.

Titrate with 0.025N thiosulphate to a pale straw colour. Add 1–2 ml of starch solution and continue the titration, till the colour disappears. Calculation if 200 ml of the original sample is titrated:

1 ml 0.0250 N Na$_2$S$_2$O$_3$ = 0.2 DO/1 ml = 1 mg DO/liter

If the thiosulphate is not 0.025 N the following formula should be used

$$\text{mg/l DO} = \frac{A \; B \; 8000}{C}$$

Where:

A = ml of Na$_2$S$_2$O$_3$
B = N of Na$_2$S$_2$O$_3$
C = ml of original sample titrated

In natural waters dissolved oxygen usually varies from near 0 to 14 ppm, the average is about 8 ppm. Wastewater from fruit processing plants generally contain very little dissolved oxygen.

Dissolved Oxygen Analyzer

The use of instruments to monitor dissolved oxygen levels in water, waste water, in fruit juices is highly recommended. The commercially available instruments are, e.g.:

− KENT Portable Digital Oxygen Meter Model 7135
− Ingold IL Model 540 Portable Dissolved Oxygen Analyzer
− Ingold IL Model 1240 In-Line Dissolved Oxygen Analyzer
− YSI Dissolved Oxygen Meter Model 58
− Orbisphere Dissolved Oxygen Analyzer Model 27165.

8.3.4 Biochemical Oxygen Demand (BOD)

The BOD, is the quantity of oxygen required for the stabilization of oxidizable organic material present after 5 days incubation at 20 C. The BOD values cannot be compared unless the results have been obtained under identical test conditions. Complete stabilization would take a much longer time than the 5-day period which has been established as an accepted standard. The extrapolation of test results to actual stream oxygen demands is highly questionable because the laboratory environment does not reproduce stream conditions such as temperature, sunlight, biological population, water movement, and oxygen concentration.

The extent of change in BOD appears to be a function of the amount of organic matter (fruit waste) and the number and types of organisms (biological population). The amount of oxygen demand in the sample will govern the need for and the degree of dilution. Samples with low DO values should be aerated to increase the initial DO content above the BOD. Let air bubble through a diffusion tube into the sample for 5 min, or until the DO is at least 7 mg/l. Determine DO on one portion of the aerated sample, seed another portion only if necessary, and incubate it for the BOD determination.

Samples for BOD analysis may undergo significant degradation during handling and storage. To reduce the change in oxygen demand that occurs between sampling and testing, keep all samples at or below 4 C and begin incubation not more than 24 h after the sample is collected.

Caustic alkalinity, mineral acid, free chlorine, and heavy metals are among the factors that may influence test accuracy. If the sample to be tested has been chlorinated, heated adulterated, and acidified, refer to the section on "Pretreatment and BOD Determination with Seeding". Although the BOD determination is generally the major test used to determine waste strength for control purposes, it has the following disadvantages in addition to those mentioned above. The BOD test is time-consuming, requires a great deal of equipment, and an inconsistent bacterial population may give unreliable results.

Apparatus:

- Glass-stoppered BOD-Incubation Bottles with collar, 300 ml capacity. Clean bottles and drain before use.
- Air incubator or water bath, thermostatically controlled at 20 C. All light should be excluded to prevent photosynthetic production of DO by algae in the sample.
- Burette, graduated to 0.1 ml, with a 50 ml capacity.
- Wide-mouthed Erlenmeyer flask, 250 ml.
- Measuring pipette, 10 ml.
- Large-tipped volumetric pipette.
- Graduated cylinder, 250 ml.

Reagents:

- All the reagents necessary for DO determination.
- *Distilled water*, water used for preparation of solutions must be from an all-glass still or must be deionized water. It must not contain copper, chlorine or organic material.
- *Phosphate buffer solution*, dissolve 8.5 g potassium dihydrogen phosphate, KH_2PO_4; 21.75 g dipotassium hydrogen phosphate, K_2HPO_4; 33.4 g disodium hydrogen phosphate heptahydrate, $Na_2HPO_4 . 7 H_2O$; 1.7 g ammonium chloride, NH_4CL in about 500 ml distilled water and dilute to 1 l. The pH of this buffer should be 7.2 without further adjustment.
- *Magnesium sulphate solution*, dissolve 22.5 g $MgSO_4 . 7 H_2O$ in distilled water and dilute to 1 l.
- *Calcium chloride solution*, dissolve 27.5 g anhydrous $CaCl_2$ in distilled water and dilute to 1 l.
- *Ferric chloride solution*, disssolve 0.25 g $FeCl_3 . 6 H_2O$ in distilled water and dilute to 1 l.
- *Seeding*, the purpose of seeding is to introduce into the sample a biological population capable of oxidizing the organic matter in the wastewater. When the sample contains very few microorganisms (as a result of chlorination, high temperature, or extreme pH) seed the dilution water. The standard seed material is settled domestic wastewater that has been stored at 20 C for 24–36 h. Use sufficient seed to produce a seed correction of at least 0.6 mg/l. Between 0.2 to 1 ml of the supernatant from settled sewage can be used per BOD bottle or 2 ml per litre of dilution water. Aqueous slurry of garden soil incubated for 24 h at room temperature can also be used. One milliliter of the supernatant can be added to each liter of dilution water. The supernatant can be preserved by freezing.

– *Preparation of dilution water*, the distilled water used should be stored at 20 C in gauze plugged bottles with an air supply for a sufficient time to become saturated with DO. If such storage is not practical, saturate the water by shaking the partially filled bottle or by aerating with a supply of clean compressed air.

Add 1 ml each of phosphate buffer, magnesium sulphate, calcium chloride and ferric chloride solutions for each liter of water. If dilution water is to be stored in the incubator, add the phosphate buffer just before using the dilution water. If the dilution water is seeded, use it the same day it is prepared.

Pretreatment:

– *Samples containing caustic alkalinity or acidity*, neutralize to pH 6.5 to 7.5 with 1 N sulphuric acid H_2SO_4 or sodium hydroxide NaOH, whereby the quantity of reagent does not dilute the sample by more than 0.5%. The pH of seeded dilution water should not be affected by the lowest sample.

– *Samples* containing residual chlorine, if the samples stand for 1–2 h, the residual chlorine will often dissipate. Destroying higher chlorine residuals in neutralized samples is carried out by adding sodium sulphite solution (Na_2SO_3). Determine the appropriate quantity of sodium sulphite solution on a 100 to 1000 ml portion of the sample by adding 10 ml of 1+1 acetic acid, followed by 10 ml potassium iodide (KI) solution (10 g/100 ml), and some drops of starch solution. If a blue colour is not produced, chlorine is absent and the BOD may be determined without decoloration. However, if a blue colour is produced, titrate the 100 ml, well mixed composite sample with 0.025 N sodium sulphite (Na_2SO_3) solution to the starch-iodide end point. To prepare the sulphite solution, dissolve 1.575 g Na_2SO_3 in 1 l distilled water. Prepare daily, this solution is not stable. The endpoint is between the last trace of blue colour and a colourless solution. Make the titration very slowly, counting the number of drops (n) of sodium sulphite solution used (or the volume).

To dechlorinate a sample for BOD testing, measure another sample and add the proportionate number of drops of sulphite solution (n) necessary for dechlorination.

– *Samples supersaturated* with dissolved oxygen, samples containing more than 9 mg/l of DO at 20 C may be encountered in cold water or in water where photosynthesis occurs (algae are growing actively). To prevent loss of oxygen during incubation of the sample, reduce the DO to saturation by bringing the sample to 20 C in a partly filled bottle and agitating it by vigorous shaking or by aerating with compressed air.

Procedure for BOD Determination Without Seeding

– Determine the pH value, residual chlorine and the sample's case history. The next step is to determine the amount of sample (S) added to the 300 ml BOD

Table 36. BOD range chart and appropriate dilutions

Expected BOD5 range, ml/l	Sample added to 300-ml bottle, ml
210–560	3
105–280	6
70–187	9
53–140	12
42–112	15
35– 94	18
30– 80	21
26– 70	24
24– 62	27
21– 56	30
14– 37	45
11– 28	60
8– 22	75
4– 12	150

bottle. To make this calculation, it should be understood that dilution water at room temperature contains approximately 8 mg/l of DO. Consequently, if the oxygen demand of the sample to be tested is greater than 8 mg/l, the sample must be diluted. It is desirable to have at least 1 mg/l of oxygen left unused after 5 days of incubation. A DO uptake of at least 2 mg/l after 5 days will produce the most reliable results. Table 36 can be used to estimate the amount of sample to add to the 300 ml BOD bottle.

− Very strong wastewater can be diluted (1 part wastewater to 9 parts dilution water) before adding 3 to 6 ml of the diluted wastewater sample into BOD bottle for analysis. This way, a range from 1000 to 3000 mg/l BOD5 can be covered.

− Note that, the initial DO is assumed to be the concentration of dissolved oxygen in mg/l of the mixture of the dilution water and the sample immediately after initial mixing.

− Fill two 300 ml BOD bottles about half full with dilution water. With a large-tipped pipette, dispense the precalculated amount of sample (S) into each of the two 300 ml bottles. Fill each bottle with dilution water and insert stoppers.

− Fill an additional two 300 ml BOD bottles with only dilution water and insert stoppers.

− Incubate at 20 C one bottle containing diluted sample and one containing only dilution water.

− Run a DO determination on the remaining BOD bottle containing the diluted sample and record the initial DO content as (D_1). Run another DO determination on the remaining BOD bottle containing the dilution water and record the initial DO content as (C_1).

− After 5 days, run a DO determination on the two incubated BOD bottles. Record the DO of the incubated diluted samples as (D_2). Record the DO content of the incubated dilution water for control as (C_2). Note that the increase or decrease of DO in bottles with just dilution water should not be used to correct the diluted sample results; it is only a measure of dilution water quality. There should *not* be an increase or decrease of more than 0.2 mg/l of DO between C_1 and C_2. Larger changes may be caused by improper testing techniques or contaminated dilution water.

– In order to store the sample after the 5 day period, fix with DO reagents and store at 4 C in the dark. Titration with 0.025 N thiosulphate can be done at a later time.
– The use of electronic oxygen monitors is highly recommended for the determination of DO.

Calculation: The following equation is used when dilution water is not seeded.

$$mg/l \; BOD_5 = \frac{100 \, (D_1 - D_2)}{P} = \frac{300 \, (D_1 - D_2)}{S}$$

Where:

D_1 = DO of diluted sample immediately after preparation, mg/l.
D_2 = DO of diluted sample after 5-day incubation, mg/l.
S = volume of sample added to the 300 ml BOD bottle, ml.
P = percentage of sample added = (S/300) 100.

Procedures for BOD Determination with Seeding

For satisfactory BOD results, an adequate number of bacteria must be present. In-plant water analysis and especially chlorinated wastewater samples requires seeding for BOD determination. After dechlorination the samples must be reseeded. To seed the samples, add 1 ml of the seed to each of the BOD bottles containing the dechlorinated sample. Also set up seed samples in three 300-ml bottles for BOD determination of the seed using 6, 9, and 12 ml seed.

Calculation: The following equations and dilution water are used when samples are both seeded.

$$mg/l \; BOD_5 \text{ of the seed} = \frac{(D_1 - D_2) \, (300)}{S}$$

Where:

D_1 = mg/l DO in seeded dilution water immediately after preparation.
D_2 = mg/l DO in seeded dilution water after 5 days incubation at 20 C.
300 = volume of BOD bottle in ml
S = ml of seed in BOD bottle

$$mg/l \; BOD_5 \text{ of seeded sample} = D_1 - D_2 - \left[\frac{(1) \, (BOD \; seed)}{300} \right] \left[\frac{300}{S} \right]$$

Where:

D_1 = initial DO of diluted and seeded sample immediately after preparation, mg/l

D_2 = 5-day DO of diluted and seeded sample after incubation, ml/l

300 = volume BOD bottle, ml

1 = ml of aged seed in BOD bottle containing sample and dilution water

S = ml of sample in BOD bottle containing 1 ml of aged seed and dilution water

Example for calculations and details of BOD determination are given by: Water Pollution Control Federation[81].

Simplified Manometric Apparatus for BOD Test

The standard five-day dilution method for BOD determination is complex and requires highly experienced and proficient personnel to get reliable results. The Hach company developed a simplified BOD procedure, in which a measured sample of wastewater and one BOD Nutrient Buffer Pillow are placed in one of the bottles on the apparatus and the bottle is connected to a closed-end mercury manometer. A quantity of air is trapped above the sample. Over a period of time, bacteria in the wastewater utilize oxygen to oxidize organic matter present in the sample and dissolved oxygen is consumed from the sample. Air in the closed sample bottle replenishes the utilized oxygen. This results in a drop of air pressure in the sample bottle. Pressure drop is registered on the mercury manometer which is calibrated directly as mg/l BOD.

The sample is agitated continually during the test period by magnetic stirrer to enhance oxygen transfer. Carbon dioxide produced by the oxidation of organic matter is removed from the system by placing a seal cup containing crystals of lithium hydroxide inside the sample bottle.

BODs from 0 to 700 mg/l can be read directly with no calculations. A continuous curve of reading gives a great deal of information about the nature of the waste.

HACH-BOD-Apparatus Model 2173 B: HACH-BOD-Nutrient Buffer Pillows contain the American Public Health Association nutrient solutions (calcium chloride, ferric chloride, magnesium sulphate) and phosphate buffer in a water-base slurry. They are packed in sterile unit-dose containers.

8.3.5 Chemical Oxygen Demand (COD)

The COD is used as a measure of the oxygen equivalent of the organic matter content of a sample that is susceptible to oxidation by a strong chemical oxidant such as potassium dichromate. The decrease in the dichromate ion after heating is measured by titration with a reducing agent. The majority of organic compounds are oxidized completely in this test.

A standard COD determination almost invariably gives higher results than a 5-day BOD determination because the oxidation of the organic material is more complete. The COD has a constant, measurable relationship to BOD when performed on fruit and vegetable waste; a ratio of 0.69 exists for the BOD/COD relationship. Actually this varies between approximately a ratio of 0.50 and 0.70 depending upon the substrate.

COD values can be determined more quickly than BOD (3 hours instead of 5 days) and can, therefore, be used to estimate plant performance much more rapidly.

The dichromate reflux method is preferred over procedures using other oxidants because of superior oxidizing activity, application to a wide variety of samples and accuracy.

In this method a sample is refluxed in strongly acid solution (sulphuric acid) with a known excess of potassium dichromate ($K_2Cr_2O_7$). After digestion the remaining unreduced $K_2Cr_2O_7$ is titrated with ferrous ammonium sulphate to determine the amount of $K_2Cr_2O_7$ consumed and the oxidizable organic matter is calculated in terms of oxygen equivalent. Keep ratios of reagent weights, volumes, and strenghts constant when sample volumes other than 20 ml are used. The standard 2-h reflux time may be reduced if it has been shown that a shorter period yields the same results.

Apparatus:

- *Reflux apparatus*, consisting of a 250-ml Erlenmeyer flask and a jacket-style condenser with 24/40 ground-glass joint.
- *Hot plate* with 1.4 W/cm^2 of heating surface.
- *Glass beads*, assorted pipettes, graduated cylinder (50-ml) and graduated burette (50-ml).

Reagents:

- Potassium dichromate solution, 0.25 N, dissolve 12.259 g $K_2Cr_2O_7$, previously dried at 103 C for 2 hours, in H_2O and make up to 1 l.
- *Ferrous ammonium sulphate solution*, 0.25 N, dissolve 98 g Fe-$SO_4(NH_4)_2SO_4$. 6 H_2O in H_2O. Add 20 ml concentrated sulphuric acid, cool and make up to 1 l. This solution must be standardized against potassium dichromate daily.
- *Sulphuric acid-silver sulphate solution*, dissolve 22 g silver sulphate (Ag_2SO_4) in a 4.1 kg bottle of concentrated sulphuric acid (H_2SO_4). It will take 1 to 2 days for the silver sulphate to dissolve.
- *Ferrion indicator solution*, dissolve 1.485 g of 1,10-phenanthroline monohydrate ($C_{12}H_8N_2$. H_2O), together with 0.695 g ferrous sulphate crystals (Fe-SO_4 . 7 H_2O) in sufficient H_2O, and dilute to 100 ml. This indicator solution may be purchased already prepared.

- *Mercuric sulphate* (HgSO₄), crystals.
- *Sulphuric acid* (H₂SO₄), concentrated 36 N
- *Sodium chloride* (NaCl), crystals.
- *Sodium hydroxide* (NaOH), solid or liquid.
- *COD Standard Solution*, 500 mg/l, lightly crush and then dry potassium acid phthalate (HOOCC₆H₄COOK) to constant weight at 120 C. Dissolve 425 mg in H₂O, and dilute to 1 l. Potassium hydrogen phthalate has a theoretical COD of 1.176 g O₂/g and this solution has a theoretical COD of 500 mg O₂/l. This solution is stable when refrigerated for up to 3 months if there is no visible biological growth.

Standardization of the Ferrous Ammonium Sulphate Solution

Pipet 10 ml 0.25 N potassium dichromate into a 250-ml Erlenmeyer flask and dilute with distilled water to about 100 ml. Add 30 ml pure concentrated sulphuric acid and allow to cool. Titrate against the ferrous ammonium sulphate using three drops of the ferrion indicator. Use the same volume of ferroin (about 0.15 ml) for all titrations. Take as the endpoint of the titration the first sharp color change from blue-green to reddish brown.

Determination of COD:

- Add a few glass beads to a 250-ml Erlenmeyer flask. Pipette a 20-ml sample in the flask, add 0.4 g mercuric sulphate and add 10.0 ml 0.25 N potassium dichromate solution. (Note: if COD is expected to be high a sample size smaller than 20 ml should be used to accomodate 10 ml potassium dichromate solution). Carefully add 30 ml sulphuric acid-silver sulphate solution, a little at a time, mixing after each addition. Be sure the reflux mixture is mixed thoroughly before heat is applied. Attach the flask to the condenser and reflux the mixture for 2 hours. Cool, and then wash down the condenser with about 25 ml distilled water.
- Dilute the mixture to about 140 ml, cool to room temperature, and titrate the excess dichromate with ferrous ammonium sulphate solution, using the ferroin indicator. Record the milliliters of ferrous ammonium sulphate solution used for the sample (B).
- A blank consisting of 20 ml distilled water instead of the sample, together with the reagents, is refluxed in the same manner. Record the milliliters of ferrous ammonium sulphate solution used for the blank (A).

Calculation:

$$mg/l \; COD = \frac{(A-B) \, C \, 8 \, 1000}{D}$$

Where:

A = Ferrous ammonium solution used for blank, ml

B = Ferrous ammonium solution used for sample, ml

C = normality of ferrous ammonium sulphate (See above)

8 = equivalent weight of O_2 . 1000 = mg/g

D = ml of sample

COD Determination Using a Space-Saving Versatile COD Reactor

The compact Hach COD dry bath reactor and reagent vials replace the hot plates, digestion flasks, and reflux condensers used in the macro COD method. The reactor is factory set to maintain 150 C. Reagent vial necks and caps reach a temperature of approximately 85 C because they extend above the reactor. The temperature differential between the top and bottom of the vial induces proper refluxing action. Samples are ready for titrimetric measurement or colorimetric determination after a two-hour digestion.

Key features include new mercury free COD reagent vials. The small reagent volumes used in Hach's semi-micro COD system not only reduces the danger from spills but also minimizes spent silver and mercury reagent disposal. Two COD reagents are available:

High range COD reagent vials (0–1500 mg/l), and low range COD reagent vials (0–150 mg/l).

The procedure involves just three steps:

– Add 2 ml of sample and recap the vial,
– Heat vials for two hours at 150 C in the Hach COD-reactor,
– Read the results directly in mg/l in a Hach spectrophotometer (at 620 nm).

The colorimetric methods are simpler and quicker than the titrimetric procedure.

9 Assessment and Improvement of Quality

9.1 Inspection, Quality Control, and Quality Assurance

As mentioned in Chapter 1 "Quality Assurance Management" the quality assurance manager should prepare his own Manual and every detail must be set up in a written form. By the use of the Manual as a guide, the function of the quality assurance department resolves itself into the periodic inspection and quality control. There is a great difference between the traditional methods of control (Inspection, and Quality Control) and Quality Assurance.

Inspection is only sorting bad from good, or testing the product for defects or for meeting the standard of a row material or a finish product. As a matter of fact, Quality cannot be inspected into a product.

Quality Control is planning for inspection operations in the cycle of supplying, production, storage and handling of a food product. This helps to identify errors, but did little to prevent them. This is what we know as "Quality Control Systems".

In a "Quality Asssurance System" problems can be anticipated and largely prevented. It is all planned and systematic actions necessary to provide confidence that a food product give satisfaction to the customer and to assure compliance with food regulations, at optimum cost to the producer. In such a system the Quality Assurance Manager has to identify the Critical Control Points (CCPs) at which quality defects can be controlled, and the quality of the final product can be improved. The Quality Assurance Department is responsible for monitoring CCPs. It must be informed of the points to monitored, the test procedures to be used, the frequency of testing, the acceptable limits, and actions to be taken when the limits are exceeded. The Quality Assurance System should include a system for record keeping so that results can be readily interpreted by the Quality Assurance Staff, production personnel and regulatory authorities. A computer monitoring programme to analyze, summarize and inform personnel responsible for control is ideal for record keeping. The Quality Assurance System is applicable at all stages in the food chain, from production of the raw material, processing, storage to ultimate use in homes and food service establishments. Quality assurance considerations and procedures are applied long before any raw material enter the plant and long after the products leave the shipping dock. Its main emphasis is on creating the conditions that lead to satisfactory performance.

9.2 Total Quality Management

The current progress in quality assurance management in general has come about through various stages over the last 60 years. During the 1930's Edward Deming and Joseph Juran developed their concepts of Quality Control (QC) and Statistical Process Control (SPC) in the United States. The concepts of Total Quality Management (TQM) have slowly been developed across the industrialized world, and especially in USA, Japan and Europe. Today several concepts are available, e.g. The Deming Managing Methods [138–142], The Juran Planning for Quality [143–144], The Ishikawa Total Quality Control [145–146], The Taguchi Methodology Whithin Total Quality [148], or Total Quality Management according to Crosby [149], Feigenbaum [150] or Okland [151].

Total Quality Management (TQM) is a way of managing an organization that aims at the continuous participation and co-operation of all in the improvement of its products, services and activities, to achieve cinstomer's satisfaction, the objectives of the organization for the benefit of all, in accordance with the requirements of society.

9.3 Standards for Quality Assurance Management Systems

The British Standards Institution (BSI) published in 1979 the general "Quality Assurance Management System" [152], which became the world Standard of the International Organization for Standardization (ISO), known as the ISO 9000 series [153]. Since the publication of the ISO 9000 series the major industrialized countries have produced their own national standards which are exact replicas of the ISO, e.g. the Deutsches Institut für Normung (DIN) [154].

The ISO Standards is produced in five parts:

– ISO 9000 Quality management and quality assurance standards: Guidelines for selection and use.
– ISO 9001 Quality systems: Model for quality assurance in design/development, production, installation and servicing.
– ISO 9002 Quality systems: Model for quality assurance in production and installation.
– ISO 9003 Quality systems: Model for quality assurance in final inspection and test.
– ISO 9004 Quality management and quality systems elements: Guidelines.

ISO 9000 is an advisory document. Its aim is twofold. Firstly, to clarify the distinctions and interrelationships among the principal quality concepts. Secondly, to provide guidelines for the selection and use of a series of Standards on quality systems that can be used for internal quality management purposes (ISO 9004), and for external quality assurance purposes in contractual situations (ISO 9001, ISO 9002, and ISO 9003). ISO 9004 is the most useful of the ISO 9000 series for a fruit processing plant. It provides detailed advice to companies on overall quality management and the quality system elements. In addition it provides guidance in other areas such as marketing, product safety and liability, and quality costs. ISO 9001 is the most complete model for quality assurance system.

It should be noted that it is not the purpose of the ISO 9000 series to standardize the quality system implemented by organizations, but to give only guidelines for selection and use.

9.4 Quality Improvement

Quality represents a condition or a set of values which has been achieved; it is usually related to something which actually exists. Perfection, however, is out of reach. Hence perfection is a good target for achievement but can not be used as measure of quality. But as an idea to be aimed at it can be made to serve practical ends for quality improvement.

According to Crosby [149] there are four absolutes for quality:

– Quality is defined as conformance to requirements, not as "goodness" or "elegance".
– The system for causing quality is prevention, not appraisal.
– The performance standard must be zero defects, not "that's close enough".
– The measurement of quality is the price of non-conformance, not indices.

Quality assurance is not just documentation and quality maintenance. It has to be a good payback for the effort involved in quality improvement. Quality improvement action depends on identifying one or more situations requiring improvement. Such situation may arise for example from a major customer problem or a major manufacture problem. It is usually the quality assurance management who make the identification and initiative action. Crosby [149] recommend the following tools to a quality improvement process:

– Establish management commitment,
– Form the quality improvement team from representatives from each department,
– Establish quality measurement throughout the company,
– Evaluate the cost of quality,
– Establish quality awareness by employees,

- Instigate corrective action,
- Establish an Ad Hoc Committee for the zero defects program,
- Supervisor/Employee training,
- Hold a zero defects day to establish the new attitude,
- Employee goal setting should set up to follow the collection of problems,
- Error cause removal should be set up to follow the collection of problems,
- Establish recognition of those who meet goals or performs outstandingly by (non-financial) aware programs.
- Quality councils composed of quality professionals and team chair persons should meet regularly, and
- Do it all over again.

According to Reinacher [155] the best way of quality assurance is to produce quality. Quality has to be present in all materials (row material, food ingredients, packaging materials etc.), technological processes, administrative processes (personal politics, purchasing, sales, marketing etc.), services and transport. Quality cannot be assured to satisfy a system or a law of a public authority, but only for the benefit of people, of the customers, of the manager themself.

Quality is not created by chance, it must be carefully planned. It is inevitable for any enterprise to define certain quality standards. Neither control nor improvement are possible without this determination of standards. Appropriate quality planning generally results in a "Total Quality Management" and involves all sections and all people of the company. If these are properly applied, it is possible to achieve a close approximation to the zero error theory (Geiss [156]).

Appendix 1

References

Chapter 2

1. Pearson D (1973) Laboratory Techniques in Food Analysis. Butterworth, London
2. Zurcher K, Hadorn (1981) Wasserbestimmung nach Karl Fischer an verschiedenen Lebensmitteln. Deutsche Lebensmittel-Rundschau 77:343
3. FAO (1988) Quality Control in Fruit and Vegetable Processing. FAO Food and Nutrition Paper 39, FAO, Rome
4. Millies K, Burkin M (1984) Anwendung der Refraktometrie zur Produktionskontrolle in Fruchtsaftbetrieben. Flüssiges Obst 51:629
5. Pearson D (1976) The Chemical Analysis of Foods. Churchill Livingstone, Edinburgh London
6. Schobinger U (1987) Frucht- und Gemüsesäfte. E Ulmer, Stuttgart
7. Prashan VC (1975) Quality Control Manual for Citrus Processing Plants. Intercit Inc, Safety Harbor, Florida
8. AOAC (1984) Official Methods of Analysis. Association of Official Analytical Chemists. Editor: Williams S, Arlington, Virginia, USA
9. Kline DA, Fernandez-Flores E, Johnson AR (1970) Quantitative determination of sugar in fruits by GLC separation of TMS derivatives. J A O A C 53:1198
10. Macrae R (1988) HPLC in Food Analysis. Academic, London, New York
11. Hurst WJ, Martin RA (1977) Rapid HPLC determination of carbohydrates in milk chocolate products. JAOAC 60:1180
12. DeVries JA, Heroff JC, Egberg DC (1979) HPLC determination of carbohydrates in food products. JAOAC 62, 1292
13. Binder H (1980) Separation of monosaccharides by HPLC. J Chromatog 189, 414
14. Damon CE, Pettitt BC (1980) HPLC determination of fructose, glucose and sucrose in molasses. JAOAC 63:476
15. Henniger G (1984) Enzymatic procedures in the analysis of fruit juices. Flüssiges Obst 51:250
16. Boehringer (1989) Methods of Biochemical Analysis and Food Analysis. Boehringer Mannheim GmbH, Sandhofer Strasse 116, D-6800 Mannheim, FR Germany
17. Bergmeyer HU (1984) UV-Methods with hexokinase and glucose-6-phosphate dehydrogenase. In: "Methods of Enzymatic Analysis" vol VI Verlag Chemie Weinheim FR Germany, p 163
18. ASU (1984) Bestimmungen von Glucose und Fructose in Fruchtsäften. Amtliche Sammlung von Untersuchungsverfahren nach § 35 LMBG, L 31.00-12, Beuth, Berlin
19. Hart FL, Fisher HJ (1971) Modern Food Analysis. Springer, Berlin, Heidelberg, New York
20. ASU (1980) Bestimmung von titrierbaren Säuren. Amtliche Sammlung von Untersuchungsverfahren nach § 35 LMBG, Beuth, Berlin
21. Baltes W (1990) Rapid Methods for Analysis of Food and Food Raw Material. Technomic, Basel, Switzerland
22. Boland Rl, Garner GB (1973) Determination of organic acids in tall fescure by GLC. J Agric Food Chem 21:661
23. Mabesa LB, Baldwin RE, Garner GB (1979) Non-volatile organic acid profiles of peas and carrots cooked by microwaves. J Food Protection 42:385
24. Lea AGH, Smith SJ (1985) HPLC in fruit juice quality control. Flüssiges Obst 52:586
25. Jeuring HJ, Brands A, Doorninck P (1979) Rapid determination of malic and citric acids in apple juice by HPLC. Z Lebensm-Untersuch Forsch 186:185

26. Bio-Rad Publication (1979) The liquid chromatography No. 2EG
27. Engelhardt H (1986) Practice of high performance liquid chromatography. Springer, Berlin Heidelberg New York
28. ASU (1984) Bestimmung von Citronensäure (Citrat) in Fruchtsäften. 31.00/14. Amtliche Sammlung von Untersuchungsverfahren nach § 35 LMBG. Beuth, Berlin
29. Mollering H (1985) Determination of citric acid. In: Bergmeyer HU (ed), Methods of enzymatic analysis, vol VII, Verlag Chemie, Weinheim, p 2
30. IFU (1985) Analytical Book of the International Federation of Fruit Juice Producers. Switzerland Obstverband, CH-8300 Zug 2
31. Wrolstad RE (1976) Color and pigment analysis in fruit products. Agriculture Experiment Station, Oregon State University, Cornwalis, USA, State Bulletin 624
32. Baker DR, Schuster R (1979) Utilization of a microprocessor based variable wavelength detector. In: Charalambous G (ed) Liquid chromatographic analysis of food and beverages vol I. Academic, New York
33. Woodward BB, Heffelfinger GP, Ruggles DJ (1979) HPLC determination of sodium saccarin, sodium benzoate and caffeine in soda beverages. JAOAC 62:1011
34. Hewlett-Packard (1981) HPLC analysis of food additives. I. Preservatives. Application Note AN 232–4
35. Vas K, Nedbalak H, Scheffer H, Kovacacs-Proszt G (1967) Methodological investigation on the determination of some pectic enzymes. Fruchtsaft Ind 12:164
36. Siliha H (1985) Studies on cloud stability of apricot nectar. Ph D Thesis, Agric University Wageningen, The Netherlands
37. Redd JB, Hendrix CM, Hendrix DL (1986) Quality Control Manual for Citrus Processing Plants. Intercit, Safety Harbor, Florida, USA
38. Handwerk RL, Coleman RL (1988) Approaches to the citrus browning problem: A Review. J Agric Food Chem 36:231
39. Klim M, Nagy S (1988) An improved method to determine nonenzymatic browning in citrus juices. J Agric Food Chem 36:1271
40. Lee HS, Nagy S (1985) Quality changes and nonenzymatic browning intermediates in grapefruit juice during storage. J Food Sci 53:168
41. Dinsmore HL, Nagy S (1974) Improved colorimetric determination for furfural in citrus juices. J A O A C 57:332
42. Marcy JF, Rouseff RL, (1984) HPLC determination of furfural in orange juice. J Agric Food Chem 32:979
43. Lee HS, Rouseff RL, Nagy S (1986) HPLC determination of furfural and 5-hydroxy-methyl-furfural in citrus juices. J Food Sci 51:1075
44. Allen B, Chin H (1980) Rapid HPLC determination of HMF in tomato paste. JAOAC 63:1074
45. Cilliers JJL, Niekerk PV (1984) Liquid chromatographic determination of HMF in fruit juices and concentrates after separation on two columns. JAOAC 67:1037
46. Mijares RM, Park GL, Nelson DB, McIver RC (1986) HPLC anaysis of HMF in orange juice. J Food Sci 51:843

Chapter 3

47. Mackinney G, Little AC (1962) Color of foods, AVI, Westport, Connecticut, USA
48. Clydesdale FM, Francis FJ (1970) Color scales. Food Prod Dev 3:117
49. DeMan JM (1985) Color. In: DeMan (ed) Principles of food chemistry. AVI, Westport, Connecticut, USA
50. VanWazer JR, Lyons JW, Kim KY, Colwell RE (1966) Viscosity and flow measurement. John Wiley, New York, USA
51. Schramm G (1990) Introduction to Practical Viscometry. Haake GmbH, Karlsruhe, FR Germany
52. Sherman P (1970) Industrial rheology. Academic, London, New York

53. Hengstenberg J, Sturm B, Winkler O (1980) Messen, Steuern und Regeln in der chemischen Technik. Springer, Berlin, Heidelberg, New York
54. Bourne MC (1982) Food texture and viscosity, concept and measurement. Academic, New York
55. Brennan JG (1980) Food Texture Measurement. In: King RD (ed) Development in analysis techniques, vol II. Applied Science, London, p 1
56. DeMan JM, Voisy PW, Rasper VF, Stanly DW (1976) Rheology and texture in food quality. AVI, Westport, Connecticut, USA
57. Klettner PG, Bielig HJ (1975) Rheologische und rheometrische Probleme bei pflanzlichen Lebensmitteln. TU Berlin, Abt. Publikationen, FR Germany
58. Kramer A, Szczesniak AS (1973) Texture measurements of foods. Reidel, Dordrecht, Netherlands
59. Rha CH (1975) Theory, determination, and control of physical properties of food materials. Reidel, Dortrecht, Netherlands
60. Sherman P (1979) Food texture and rheology. Academic, New York
61. The Journal of Texture Studies. Food and Nutrition Press, 1 Trinty Square, Westport, Connecticut 06880, USA
62. Committee of Pectin Stantardization, Institute of Food Technologists (1959) Final report. Food Technol 13:396
63. Troller JA (1980) Influence of water activity on microorganisms in foods. Food Technol 34 May, 76
64. Troller JA (1983) Methods of measure water activity. J Food Protection 46:129
65. Stoloff L (1978) Calibration of water activity measuring instruments and devices: Collaborative study. JAOAC 61:1166
66. Troller JA, Christian JHB (1978) Water activity and food. Academic, London

Chapter 4

67. DIN and ISO. Standard methods. Beuth, Berlin, FRG
68. Jellinek G (1981) Sensory evaluation of food. Siegfried, Pattensen, FR Germany
69. Gridgeman NT (1984) Testing Panels: Sensory Assessment in Quality Control. In: Herschdörfer SM (ed) Quality control in the food industries, vol I. Academic, London
70. Moskowitz H (1988) Applied sensory analysis of foods. vols I, II, CRC Press, Boca Raton, FL, USA
71. Piggott JR (1984) Sensory analysis of food. Elsevier, London
72. Hubbard MR (1990) Statistical quality control for the food industry. Van Nostrand Reinhold, New York
73. Munoz ZA, Civille GV, Carr BT (1992) Sensory evaluation in quality control. Van Nostrand Reinhold, New York
74. Roessler EB, Pangborn RM, Sidel JL, Stone H (1978) Expanded statistical tables for estimating significance in paired-preference, paired-difference, duo-trio and triangle test. J Food Sci 43:940
75. Kahan F, Cooper D, Papavasiliou A, Karmer A (1973) Expanded tables for determining significance of differences for ranked data. Food Technol 27:61
76. Merrington M, Thompson CM (1943) Tables of percentage points of the inverted beta (F) distribution. Biometrika 33:73
77. Duncan DB (1955) Multiple range and multiple F-Tests. Biometrics 11:1

Chapter 5

78. Pichhardt K (1989) Lebensmittelmikrobiologie. Springer, Berlin Heidelberg New York
79. Fields ML (1979) Fundamentals of Food Microbiology. AVI, Westport, Connecticut, USA

80. ICMSF (1980) Microbiological ecology of foods. International commission on microbiological specification for foods. Academic, London
81. Jay JM (1984) Modern Food Microbiology. Van Nostrand, New York
82. Kramer J (1987) Lebensmittelmikrobiologie. Eugen Ulmer, Stuttgart, FR Germany
83. Müller G, Lietz P, Munch HD (1987) Mikrobiologie pflanzlicher Lebensmittel. Steinkopf, Darmstadt, FR Germany
84. Panezai AK (1978) Microbiology. In: Green LF (ed) Developments in Soft Drinks Technology I. Elsevier Applied Science, London, chap 9
85. Batchelor VJ (1984) Further microbiology of soft drinks. In: Houghton HW (ed) Developments in Soft Drinks Technology III. Elsevier Applied Science, London, chap 5
86. Speck ML (1984) Compendium of methods for the microbiological examination of foods. American Public Heath Association, Washington DC
87. Buchanan RE, Gibbons NE (1970) Bergey's manual of determinative bacteriology. Williams and Wilkins, Baltimore, USA
88. Faddin JW (1977) Biochemical tests for identification of medical bacteria. Williams and Wilkins, Baltimore, USA
89. Lodder J (1970) The yeasts. A taxonomic study. North Holland, Amsterdam
90. Taha RA, Askar A, Omran HT, Mahgoub SS (1990) Microbial stability of mango soft drink. Flüssiges Obst 57:726
91. Hill EC, Wenzel FW, Barreto A (1954) Colorimetric method of detection of microbiological spoilage in citrus juices. Food Technol 8:168
92. Parish ME, Braddock RJ, Wicker L (1990) Gas chromatographic detection of diacetyl in orange juice. J Food Quality 13, 249
93. Ackland MR, Reeder JE (1984) A rapid chemical spot test for the detection of lactic acid as an indicator of microbial spoilage in processed foods. J Applied Bacteriol 56:415
94. Noll F (1984) Determination of lactic acid. In "Methods of Enzymatic Anaysis", Bergmeyer HU Vol VI pp 582, Verlag Chemie, Weinheim, FR Germany
95. Gawehn K (1984) Determination of lactic acid. In "Methods of Enzymatic Analysis", Bergmeyer HU Editor, Vol VI pp 588. Verlag Chemie, Weinheim, FR Germany

Chapter 6

96. Trinkwasserverordnung (1990) von 5. Dezember 1990, BGBl. I S 2612, Ber. 23.1.1991. FR Germany
97. APHA (1985) Standard Method for the Examination of Water and Wastewater. American Public Health Association. 1015 Eighteenth street NW, Washington DC 20036, USA
98. Holl K, Carlson S (1979) Water. Walter de Gruyter, Berlin
99. Jems GV (1971) Water Treatment. Technical Press, London
100. Matz SA (1965) Water in Foods. AVI Publishing Company Inc, Westport, Connecticut
101. McCoy JW (1969) Chemical Analysis of Industrial Water. MacDonald Technical & Scientific, London
102. McCoy JW (1981) The Chemical Treatment of Boiler Water. Chemical Publishing Co., New York
103. Minear RA, Keith LH (1982) Water Analysis Vols I, II and III, Academic, New York and London
104. Taylor EW (1959) The Examination of Water and Water Supplies. JA Churchill, London
105. MERCK E (1987) Rapid Test Handbook. MERCK Publication, D-6100 Darmstadt

Chapter 7

106. American Iron and Steel Institute (1970) Composition of some Stainless Steels. New York
107. HENKEL (1990) Detergents for fruit juice plants. P3 Division, Henkel AG, P.O.Box 1100, D-4000 Düsseldorf 1

108. Cook DJ, Binsted R (1975) Food Processing Hygiene. Food Trade Press, London
109. Elliott RP (1980) The Microbiology of Sanitation (Chapter VI) and Cleaning and Sanitation (Chapter VII). IN "Principles of Food Processing Sanitation", Editor: Denny B. Food Processors Institute, Washington DC
110. Fox A (1971) Hygiene and Food Production. Churchill Livingstone, Edinburgh London
111. Guthrie RK (1988) Food Sanitation. Van Nostrand Reinhold, New York
112. Jowitt R (1980) Hygienic Disign and Operation of Food Plants. AVI Publishing Company, Westport, Connecticut
113. Litsky BY (1973) Food Service Sanitation. Modern Hospital Press, Chicago
114. Silliker JH (1980) Microbial Ecology of Foods. Vol I, Academic, New York
115. Cornwell PB (1976) The Cockroach. Assoc Business Programmes, London
116. Davis RA (1970) Control of Rat and Mice.Ministry of Agriculture, Fisheries and Food. Bulletin 181, H M S O, London
117. Dusvine JR (1980) Insects and Hygiene. Chapman and Hall, London
118. Goldenberg N, Rand C (1971) Rodent Control in the Food Industry, Manufacture, Distribution and Retailing. Proc 3rd Br Pest Control Conf, Jersey Oct 5–8, 151–156
119. Greaves JH (1982) Rodent Control in Agriculture. FAO, Production and Protection, Paper 40, FAO Rome
120. Meehan AP (1984) Rats and Mice, Their Biology and Control. Rentokil Limited, East Grinstead, UK
121. Ministry of Agriculture, Fisheries and Food, London
    (1976) Cockroaches. Advis Leaflet no 383
    (1977) Ants Indoors. Advis Leaflet no 366
    (1977) Pests in Food Stores. Advis Leaflet no 483
122. Shenker AM (1971) Pest Control in the Food Industry. In "Hygiene and Food Production, Editor: Fox A. Churchill Livingstone, Edinburgh
123. World Health Organization (1972) Cockroaches, Biology and Control. WHO/VBC 72.354:1

Chapter 8

124. Passavant-Werke (1990) Treatment of Waste Water. D-6209 Aarbergen 7
125. Bureau of National Affairs (1979) US National Discharge Standards, USA
126. Brunner L, Bauer H (1976) Abwasseruntersuchungen in Fruchtsaftbetrieben nach dem Abwasserabgabengesetzentwurf und zu erwartende finanzielle Belastung. Flüssiges Obst 43:79
127. Farral WA (1979) Food Engineering Systems. Vol II Utilities, Chapter 11 Waste Disposal, by Rippen AL, Farrall AW. AVI Publishing Company, Textbook Series, Westport, Connecticut
128. Fresenius W, Schneider W (1989) Waste Water Technology. Springer, Berlin Heidelberg New York Paris
129. Georgia Institute of Technology (1989) Food Processing Waste Conference. Education Extension, Atlanta, Georgia, USA
130. Green JH, Kramer A (1980) Food Processing Waste Magagement. AVI Publishing Company, Westport, Connecticut
131. Jappelt W (1979) Abwasser und Abwasserreinigung bei der Obst- und Gemüseverwertungsindustrie. Hannoversche Industrieabwassertagung. Institut für Siedlungswasserwirtschaft, Universität Hannover
132. Jones HR (1973) Waste Disposal Control in the Fruit and Vegetable Industry. Noyes Data Corporation, Park Ridge, New Jersey
133. Rosenwinkel KH (1988) Saving by own waste water treatment measures. Confructa 31:106
134. Rüffer HM, Rosenwinkel KH (1981) Waste Water of the Food Industry. International Symposium on Management of Industrial Wastewater in the Developing Nation. High Institute of Public Health, 165 El-Horria Av Alexandria, Egypt March 28–31
135. Rüffer HM, Rosenwinkel KH (1986) Abwasser in der Fruchtsaftindustrie. Getränkeindustrie 11:1007

136. Water Pollution Control Federation (1985) Simplified Laboratory Procedures for Wastewater Examination. 2626 Pennsylvania Avenue NW, Washington DC 20037
137. Yahaskel A (1979) Industrial Waste Water Cleanup. Noyes Data Corporation, Park Ridge, New Jersey

Chapter 9

138. Deming WE (1982) Quality, Productivity, and Competitive Position. MIT Center for Advanced Engineering Study, Cambridge, Mass, USA
139. Mann NR (1985) The Key to Excellence: The Story of Deming Philosophy, Prestwick Books, Los Angeles, USA
140. Scherkenback WW (1986) The Deming Route to Quality and Productivity, Road Maps and Road Blocks. CEE Press Books, Washington DC, USA
141. Walton M (1989) The Deming Management Method. Mercury Books, London, England
142. Gitlow HS, and Gitlow SJ (1987) The deming Guide to Quality and Competitive Position. Prentice-Hall Inc, Englewood Cliffs, NJ, USA
143. Juran JM (1975) Quality Control Handbook. McGraw-Hill, New York, USA
144. Juran JM (1988) Juran on Planning for Quality. The Free Press, New York, USA
145. Ishikawa K (1982) Guide to Quality Control. Asian Productivity Organization, Tokyo, Japan
146. Ishikawa K (1985) What is Total Quality Control? The Japanese way, Prentice-Hall, London, England
147. Bendell A (1990) Taguchi Methodology Within Total Quality. IFS Publications, Bedford, England
148. Taguchi G (1986) Introduction to Quality Engineering, Asian Productivity Organization, Tokyo, Japan
149. Crosby PB (1984) Quality without Tears, McGraw-Hill, London, England
150. Feigenbaum AV (1983) Total Quality Control. McGraw-Hill, London, England
151. Oakland JS (1989) Total Quality Management. Heinemann, Oxford, England
152. British Standard Institute (BS) BS 5750 A guide to Quality Assurance. 2 Park Street, London W1A 2BS
153. International Organization for Standardization (ISO). Case Postale 56, CH-12211 Geneve 20, Switzerland
154. Deutsches Institut für Normung e.V. (DIN)
    DIN 55 350 Teil 11 Grundbegriffe der Qualitätssicherung.
    DIN 55 350 Teil 16 Begriffe zu Qualitätssicherungssystemen.
    DIN 55 350 Teil 17 Begriffe der Qualitätsprüfungsarten.
    DIN ISO 9000 Leitfaden zur Auswahl und Anwendung der Normen zu Qualitätsmanagement, Elementen eines Qualitätssicherungssystems und zu Qualitätssicherungs-Nachweisstufen.
    DIS ISO 9001 Qualitätssicherungs-Nachweisstufen für Entwicklung und Konstruktion, Produktion, Montage und Kundendienst.
    DIN ISO 9002 Qualitätssicherungssysteme;
    Qualitätssicherungs-Nachweisstufe für Produktion und Montage.
    DIN ISO 9003 Qualitätssicherungssysteme:
    Qualitätssicherungs-Nachweisstufe für Endprüfungen.
    DIN ISO 9004 Qualitätsmanagement und Elemente eines Qualitätssicherungssystems
    Deutsches Institut für Normung e.V., Beuth-Verlag, Burggrafenstrasse 6, 1000 Berlin 30
155. Reinacher E (1991) Practice-oriented and consistent quality assurance. Flüssiges Obst – Fruit Processing 1,(6), 104
156. Geiss H (1992) Quality assurance and quality management in the fruit juice industry. Flüssiges Obst 59, 130

Recommended Reading

Books

AFIS (1952) Sanitation for the Food Preservative Industries. Association of Food Industry Sanitations. McGraw Hill, New York

Amerine MA, Pangborn RM, Roessler EB (1965) Principles of Sensory Evaluation of Foods. Academic, New York

AOAC (1984) Official Methods of Analysis of the Association of Official Analytical Chemists. Editor: Williams S, Arlington, Virginia

APHA (1985) Standard Methods for the Examination of Water and Wastewater. American Public Health Association. 1015 Eighteenth street NW, Washington DC 20036

Askar A (1987) Verarbeitung von Tropischen Früchten. In "Frucht- und Gemüsesäfte". Editor: Schobinger U, E Ulmer, Stuttgart, FR Germany

Bourne MC (1983) Special Problems that Face Food Science in a Developing Tropical Country. Canadian Institut of Science and Technology Journal, Vol 6: A84

DeMan JM, Voisy PW, Rasper VF, Stanley DW (1980) Rheology and Texture in Food Quality. AVI Publishing, Westport, Connecticut

Gould WH (1977) Food Quality Assurance, AVI Publishing, Westport, Connecticut

Green JH, Kramer A (1980) Food Processing Waste Management. AVI Publishing, Westport, Connecticut, USA

Guthrie PK (1988) Food Sanitation. AVI Publishing, Westport, Connecticut. USA.

Herschdoerfer SM (1984/1986) Quality Control in the Food Industry. Vols I, II and III. Academic, London and New York

IFU (1985) Analytical Book of the International Federation of Fruit Juice Producers. Schweiz. Obstverband, CH-6300 Zug 2

Jagtiani J, Chan HT, Sakai WS (1988) Tropical Fruit Processing. Academic, London

Jellinek G (1981) Sensory Evaluation of Food. Verlag D&PS, Pattensen, FR Germany

Jones HR (1973) Waste Disposal Control in the Fruit and Vegetable Industry. Noyes Data Corporation, New Jersey

Koch J (1986) Getränkebeurteilung. E Ulmer, Stuttgart FR Germany

Kramer A, Twigg BA (1974) Quality Control for the Food Industry. Vols I and II. AVI Publishing, Westport, Connecticut

MacLeod AJ (1973) Instrumental Methods of Food Analysis. Elek Science Publisher, London

Macrae R (1988) HPLC in Food Analysis. Academic, London, New York

Meloan, CE, Pomeranz Y (1977) Food Analysis Laboratory Experments. AVI Publishing, Westport, Connecticut

Moskowitz H (1988) Applied Sensory Analysis of Foods. Vols I and II. CRC Press, Boca Raton, Florida

Nagy S, Shaw PE (1980) Tropical and Subtropical Fruits. AVI Publishing, Westport, Connecticut

Nagy S, Shaw PE, Veldhuis MK (1977) Citrus Science and Technology. Vols I and II. AVI Publishing, Westport, Connecticut

Parker ME, Litchfield JH (1962) Food Plant Sanitation. Reinhold, New York

Pearson D (1973) Laboratory Techniques in Food Analysis. Butterworth, London

Pearson D (1976) The Chemical Analysis of Foods. Churchill Livingstone, Edinburgh

Piggott JR (1984) Sensory Analysis of Food. Elsevier, London

Pomeranz Y, Meloan CE (1971) Food Analysis: Theory and Practice. AVI Publishing, Westport, Connecticut

Prashan VC (1970) Quality Control Manual for Citrus Processing Plants. Intercit Inc, Safety Harbor, Florida

Ranganna S (1986) Handbook of Analysis and Quality Control for Fruit and Vegetable Products. Tata McGraw Hill, New Dehli, India

Schobinger U (1987) Frucht und Gemüsesäfte. E Ulmer, Stuttgart, FR Germany

Speck ML (1984) Compendium of Methods for the Microbiological Examination of Foods. American Public Health Association, Washington DC
Tanner H, Brunner HR (1979) Getränke Analytik, Heller, Schwäbisch Hall, FR Germany
Woodruff JG, Luh BS (1975) Commercial Fruit Processing, AVI Publishing, Westport, Connecticut

Journals

Confructa Studien (German/English)

Flüssiges Obst GmbH, Diezerstr. 7, Postfach 1

D-5429 Schönborn, FR Germany

Flüssiges Obst (German/English)

Diezerstr. 7, Postfach 1

D-5429 Schönborn, FR Germany

Food Chemistry

Elsevier Applied Science Publishers

Crown House, Linton Road, Barking

UK-Essex IG 11 8JU

Food Technology

Institute of Food Technologists,

221N. La Salle street, Chicago

Illinois 60601, USA

Indian Food Packer

Indian Food Preserver's Association,

New Dehli, India

Journal of Agriculture and Food Chemistry

American Chemical Society,

1155 Sixteenthstreet NW,

Washington DC 20036, USA

Journal of Association of Official Analytical Chemists

Association of Official Analytical Chemists,

Suite 400, 2200 Wilson Blvd,

Arlington, UA 22201–3301, USA

Journal of Food Protection

International Association of Milk, Food and Environmental

Sanitation Inc, 502 L. Lincoln Way, P.O.Box 701

Ames, IA 50010–0701, USA

Journal of Food Science

Institute of Food Technologists,

221N. La Salle street, Chicago

Illinois 60601, USA

Appendix 2

International Organisations

CODEX
Joint FAO/WHO Food Standard Programme
FAO, Via della Terme di Caracalla,
I-00100 Rom, Italy

IFU
The International Federation of Fruit Juice Producers
10, rue de Liege,
F-75009 Paris 9e, France

ISO
The International Organization for Standardization
1, rue de Varembe, Case Postale 56,
CH-1211 Geneve 20, Switzerland

FDA
Food and Drug Administration
Division of Food Technology
200C Street SW,
Washington DC 20204, USA

USDA
United States Department of Agriculture
Fruit and Vegetable Division,
Processed Products Branch,
Room 0709, South Building,
Washington DC 20250, USA

Page	Substances/Equipment	Type/Brand	Company and Address
14	Enzymes		Novo Ferment Vogesenstr. 132 CH-4013 Basle
			Röhm GmbH Kirchenallee D-6100 Darmstadt
37	Partisil		Technicol High Performance Columns, Book street, Higher Hillgate Stockport, Cheshire SK 1 3HS, UK
58	Colour Meter	PANTONE Colour	Guide Pantone Inc, 55 Knickerbocker Road Moonachie New Jersey 07074, USA
		Hunterlab	Hunter Association Laboratory Inc. 11495 Sunset Hills Road Reston, Virginia 22090, USA
		Minolta	30, 2-Chrome, Azuchi- Machi, Higashi- Ku Osaka 541, Japan
64	Capillary Viscometer	Ostwald Cannon Ubbelohde	JENA'er Glaswerk, Schott Gen., Hattenbergstr. 10 D-6500 Mainz
		AOAC	SMI Inc, 800 University Ave, Berkeley, CA 94710 USA
67	Rotational Viscometer	Brookfield	Brookfield Engineering Laboratories Inc. 240 Cushing street Stoughton, Massachusett 02072, USA
		Rotovisco Viscotester	HAAKE Mess-Technik GmbH Co., Dieselstr. 4 D-7500 Karlsruhe 41
		Viscotron	Brabender Measurement & Control Systems, Kulturstr. 51–55 D-4100 Duisburg 1

Page	Substances/Equipment	Type/Brand	Company and Address
70	Distance Consistometer	Bostwick	Fissher Scientific, 113 Hartwell Ave. Lexington MA 02173, USA
72	Hand-Operated Fruit Firmnes Testers	Chatillon	John Chatillon & Sons Inc. Force Measurement Division, 7609 Business Park Drive, Greensboro, North Carolina 27409 USA
		Effi-GI	ITALTEST di PETRONICI PIECARLA Via Reale 63, I-48011 Alfosine, Italy
74	Mechanical and Motorized	Bloom Gelometer	CGA Precision Scientific Group 3737 W Cartlandstreet Chicago, IL 60647, USA
		Stevens LFRA	C Stevens and Sons Ltd. 2–8 Dolphin Yard Holywell Hill, St. Albans Hertfordshire AL1 1EX UK
75	Distance Measuring Instruments	Ridgelimeter	Sunkist Growers Inc. Products Sales Division, Ontario CA 91764, USA
		SUR Penetrometer	Sommer und Runge AG Bennigsenstr. 23 D-1000 Berlin 41
79	Multiple Texture	Instron	Instron Ltd, Coronation Road, High Wycome Bucks HP12 3SY, UK
		Kramer	Food Technology Corporation 123000 Parklawn Drive Rockville, MD 20852 USA
		General Foods	Zenken Company Ltd, Kyod Bldg, No. 5, 2-Chrome Honcho, Nihonbashi Chuo-Ku, Tokyo 103, Japan

Page	Substances/Equipment	Type/Brand	Company and Address
81	Water Activity Measuring Instruments	Rotronic	Rotronic AG Badenerstr. 435 CH-8040 Zürich
		Novosina	Novosina AG Thurgauerstr. 50 CH-8050 Zürich
111	Media and Ingredients for Microbiological Analysis		Aldrich Chemicals Co. Inc. 940 W Saint Paul Avenue Milwaukee, WI 52233, USA BBL, Division of Becton and Co. P.O. Box 243, Cockeysville MD 21030, USA
			DIFCO Laboratories Detroit, MI 48201, USA
			Fluka Feinchemikalien GmbH Lilienthalstrasse 8 D-7910 Neu-Ulm
			Key Scientific Products Key Substrates P.O. Box 66307 Los Angeles, CA 90066, USA
			Merck E Frankfurter Strasse 250 D-6100 Darmstadt 1
			Millipore Corporation Bedford, P.O. Box 255 MA 01730, USA
			Oxoid Limited, Wade Road Basingstoke, Hampshire RG 24 OPW, UK
			Pfizer Diagnostics 300 W 43rd street New York, NY 10036, USA
			Roche Diagnostics Hoffmann-La Roche CH-4002 Basle
			Sartorius GmbH Weender Landstrasse 94–108 D-3400 Göttingen

Page	Substances/Equipment	Type/Brand	Company and Address
			Sigma Chemicals Co. P.O. Box 14508 St. Louis, MO 63178, USA
126	Kits and Instruments for Water Analysis		Fisher Scientific 1600 Parkway View Drive, Pittsburgh Pennsylvania 15205, USA
			Hach Company, P.O. Box 389 Loveland, Colorado 80539 USA
			Merck E, Postfach 4119 D-6100 Darmstadt.
			Taylor Chemicals Inc. 7300 York Road, Baltimore MD 212204, USA
			Cole Parmer Instrument Company 7425 North Oak Park Avenue Chicago, Illinois 60648, USA
			Kleinfeld Labortechnik Leisewitzstr. 47 D-3000 Hannover 1
			Ciba Corning Analytical Colcherter Road, Essex C09 2DX, UK.
			British Drug Houses Ltd Broom Road, Poole BH12 4NN, Dorset, UK
146	UV-Treatment	HANOVA	Hanova Ltd, 145 Farnham Road, Slough Berkshire, SL1 4XB, UK
166	Detergents for Fruit Processing Plants	Henkel	Henkel, P3 Division P.O. Box 1100 D-4000 Düsseldorf 1
173	Fruit Processing Plant	Euro Citrus	Euro Citrus, P.O. Box 227 49 AE Oosterhout (NB), NL
177	Equipment for Air Analysis		Millipore Corporation Bedford, P.O. Box 255 Ma 01730, USA

Page	Substances/Equipment	Type/Brand	Company and Address
			Sartorius GmbH Postfach 3243 D-3400 Göttingen
			BGI Incorporated 58 Guinan street Waltham, MA 02154, USA
			New Brunswick Scientific Company, P.O. Box 986 44 Talmadge Road, Edison NJ 08817, USA
180	Insect and Rodent Control	DEKUR	DEKUR, Postfach 508 D-5400 Koblenz
194	Waste Water Treatment	Passavant	Passavant-Werke AG D-6209 Aarbergen
		Jost	Jost GmbH, Hammerstrasse 95 D-4400 Münster
		Westfalia	Westfalia Separator AG Postfach 3720 D-4740 Oelde
		Alfa-Laval	Alfa-Laval AB S-14780 Tumba
		Flottweg	Flottweg Werk Postfach 1160 D-8313 Vilsbiburg
		Sharples	Sharples-Stokes Division Pennwalt-Corporation 955 Mearns Road Warminster, PA 18974, USA
		Aqua Consult	Aqua Consult Ingenieur GmbH, Lange Laube 29 D-3000 Hannover 1
		Sulzer	Sulzer Chemtech. Gebrüder Sulzer AG CH-8401 Winterthur
199	Waste Water Analysis		Winopal Forschungbedarf Echternfeld 25 D-3000 Hannover

Page	Substances/Equipment	Type/Brand	Company and Address
			Fisher Scientific 711 Forbes Avenue Pittsburgh, PA 15219 USA
			Obrisphere GmbH Ludwigsstrasse 35 D-6300 Giessen
			Obrisphere GmbH, 114 Route de Thonon CH-1222 Vesenaz Geneva
			Yellow Spring Instruments. Yellow Springs, Ohio 45387 USA
			Kent Industrial Measurements, Oldends Lane, Stonehouse, Gloucestershire GL10 3TA, UK
			Hach, P.O. Box 389 Loveland, CO 80539 USA
			Hach, P.O. Box 229 B-5000 Namur 1

Subject Index

B. Wenclawiak (Ed.)

Analysis with Supercritical Fluids: Extraction and Chromatography

1992. XIV, 213 pp. 117 figs. (Springer Laboratory)
Hardcover DM 154,– ISBN 3-540-55420-3

The editor has brought together in a compact and readable form the new methods of analytical chemistry using supercritical fluids. The volume provides comprehensive treatment of supercritical fluid chromatography (SFC) and supercritical fluid extraction (SFE); it discusses both theory and practice. The contributions are written by leading experts in their fields with exhaustive practical experience in SFC and SFE. Special attention is given to the description of applications and thus provides the experienced analyst with invaluable information widely scattered in the literature and helps the novice to adopt these new techniques quickly in his laboratory. An addendum includes brand new literature on SFC and SFE.

Price is subject to change without notice. All prices for books and journals include 7% VAT. In EC countries the local VAT is effective.